SpringerBriefs in Applied Sciences and Technology

SpringerBriefs present concise summaries of cutting-edge research and practical applications across a wide spectrum of fields. Featuring compact volumes of 50 to 125 pages, the series covers a range of content from professional to academic.

Typical publications can be:

- A timely report of state-of-the art methods
- An introduction to or a manual for the application of mathematical or computer techniques
- A bridge between new research results, as published in journal articles
- A snapshot of a hot or emerging topic
- An in-depth case study
- A presentation of core concepts that students must understand in order to make independent contributions

SpringerBriefs are characterized by fast, global electronic dissemination, standard publishing contracts, standardized manuscript preparation and formatting guidelines, and expedited production schedules.

On the one hand, **SpringerBriefs in Applied Sciences and Technology** are devoted to the publication of fundamentals and applications within the different classical engineering disciplines as well as in interdisciplinary fields that recently emerged between these areas. On the other hand, as the boundary separating fundamental research and applied technology is more and more dissolving, this series is particularly open to trans-disciplinary topics between fundamental science and engineering.

Indexed by EI-Compendex, SCOPUS and Springerlink.

Lizhou Wu · Jianting Zhou

Rainfall Infiltration in Unsaturated Soil Slope Failure

 Springer

Lizhou Wu
Chongqing Jiaotong University
Chongqing, China

Jianting Zhou
Chongqing Jiaotong University
Chongqing, China

ISSN 2191-530X   ISSN 2191-5318 (electronic)
SpringerBriefs in Applied Sciences and Technology
ISBN 978-981-19-9736-5   ISBN 978-981-19-9737-2 (eBook)
https://doi.org/10.1007/978-981-19-9737-2

This Springer imprint is published by the registered company Springer Nature Singapore Pte Ltd.
The registered company address is: 152 Beach Road, #21-01/04 Gateway East, Singapore 189721, Singapore

Foreword by Jianbing Peng

It is a great pleasure to write this Foreword for Dr. Lizhou Wu's book on *Rainfall Infiltration in Unsaturated Soil Slope Failure*. I have ever taken a part in guiding the research and enjoyed our discussions on rainfall-caused landslides.

Rainfall-induced landslides pose hazards to human safety and economic development in the world. The complex geological environments and variable precipitation lead to difficulties in the assessments of slope stabilities. Lizhou's research group has performed a lot of fundamental researches on the scientific problems of rainfall-induced slope failure.

This book by Lizhou Wu and Jianting Zhou bring many creative studies on rainfall seepage and slope failure mechanisms using both numerical techniques and analytical methods. The authors are particularly successful in addressing the analytical and numerical solutions of linear and nonlinear infiltrations related to the investigation of rainfall-induced unsaturated soil slope failures. It proposes creative novel methods for investigating the linear and nonlinear systems of rainfall infiltration inducing unsaturated soil slope failure. The analytical solutions to water infiltration consider vegetation root and coupling effects, which deserve further research. These methods in the book are applicable not only to unsaturated infiltration issues, but also to rainfall-caused landslides.

I am pleased to witness the publication of this book. Students, engineers, and researchers who focus on the unsaturated infiltration and slope stability can benefit a lot from this book. I am confident that this book will serve as an important source of reference for future studies on rainfall-caused landslides.

Jianbing Peng

Jianbing Peng
Academician of Chinese Academy
of Sciences
Chang'an University
Xi'an, China

Foreword by Huiming Tang

The book titled *Rainfall Infiltration in Unsaturated Soil Slope Failure* by Lizhou Wu and Jianting Zhou is a welcomed addition that provides an engineering methodology for a challenging problem of interest the world over. Virtually, many countries encounter serious geological hazards including rainfall-triggered landslides.

The book brings primary application areas of engineering which have been individually successful within civil engineering, namely the modeling of rainfall infiltration and rainfall-induced landslides. The authors are well aware of the extensive research that has been undertaken in various parts of the world related to modeling rainfall infiltration. The development of numerical and analytical methods of rainfall seepage and slope stability is particularly well presented in this book. The authors are particularly successful in addressing the many components related to the assessment of rainfall-induced landslides.

When addressing issues related to modeling rainfall-induced landslides, there is a need to understand reliable numerical solutions of the nonlinear seepage equations. The authors have also made a significant contribution in developing numerical methods related to the estimation of unsaturated seepage for investigating rainfall-induced landslides. I believe that this book will also form an important reference book that will be in demand in university and other engineering libraries.

Tang Hui

Huiming Tang
Chair Professor, China University
of Geosciences
Wuhan, China

Foreword by Limin Zhang

Rainfall-triggered landslides are a common disaster around the world. Their mechanisms are still not well understood and their evaluating methods not well developed. Rainfall infiltration is particularly difficult to assess when soil slopes are in complex geological environments. Many methods including analytical and numerical solutions have been developed to predict slope instability due to rainfall infiltration. The predicted slope performance may deviate from the reality because of neglecting complex infiltration processes and geoenvironments. Due to the high nonlinearity of the equations governing rainfall infiltration into slopes and the hydraulic property functions, the efficient and reliable solution of the infiltration equations related to rainfall-triggered landslides is a difficult task.

This book by Lizhou Wu and Jianting Zhou combines the complex topics of analytical and numerical modeling of rainfall infiltration and seepage into unsaturated soil slopes along with the analysis of slope stability. Particular attention is given to understanding nonlinear unsaturated infiltration process. The authors have brought the many components related to rainfall-induced landslides in one nutshell. The authors have also made a significant contribution to the development of numerical and analytical methods related to the evaluation of unsaturated infiltration into soil slopes for stability analysis purposes. I am confident that this book will be an important reference to researchers, graduate students, and practicing engineers.

Limin Zhang
Chair Professor, Head of Department
of Civil and Environmental
Engineering
The Hong Kong University of Science
and Technology
Hong Kong, China

Preface

Landslides pose immediate threats to the infrastructures such as buildings, roads, bridges, and geoenvironment in the mountainous areas. Rainfall is the primary trigger of landslides that frequently cause fatalities and large economic loss. Rainfall-induced landslides gain greater attentions as climate becomes more extreme.

A number of complicated mechanisms of rainfall-caused landslides are involved in the analysis of slope stability due to rainfall. Studies on rainfall-induced landslides require good knowledge of not only mechanical properties of the soil, but also hydrologic behavior of the soil governed by the soil seepage properties.

The primary objective of the book clearly presents rainfall infiltration and slope stability analysis methods using analytical and numerical approaches. Analytical solution can consider coupled infiltration and deformation in unsaturated soil slopes considering vegetable root. A series of improved linear or nonlinear iterative methods are developed to solve complex nonlinear infiltration equation, which improves the convergence rate, accuracy, and stability. These methods are applied to simulate rainfall infiltration-induced unsaturated soil slope failures.

The improved numerical methods, nonlinear and linear iterative methods which can be used to address the related unsaturated infiltration problems are also presented. This book is an essential reading for researchers and graduate students who are interested in rainfall infiltration, slope stability, landslides, and geohazards in the fields of civil engineering, engineering geology, and earth science. The book is written to guide professional engineers and practitioners in slope engineering and geohazard management. This book can enhance their understanding of rainfall-induced landslides, help them analyze a specific problem, and prevent landslides and design engineering slopes according to the local soil and climate conditions.

Lizhou Wu
Jianting Zhou
Chongqing Jiaotong University
Chongqing, China

Acknowledgements The authors would like to express the special appreciation and respect to Prof. Jianbing Peng, who gives valuable guidance, constructive suggestions, and continued encouragement throughout this research work. The first author is grateful to his supervisors, Prof. Runqiu Huang and Prof. Limin Zhang, for years of supervision and help. The first author also extends thanks to Prof. Hong Zhang and Prof. Jun Yang for their encouragement and help.

The authors are grateful to Dr. Shuairun Zhu for his contribution to Chaps. 3 and 4, and the editing help provided by Mr. S. H. Li, Mr. Bo He, Mr. Hao Li, and Mr. Tao Ma.

The support of the National Natural Foundation of China (Nos. U20A20314, 41790440, 41790432, and 42277183), Natural Science Foundation Innovation and Development Foundation of Chongqing (No. CSTB2022NSCQ-LZX0044), and the Natural Science Fund for Distinguished Young Scholars of Chongqing (cstc2020jcyj-jqX0006) are acknowledged.

The original version of the book was revised: Funder information has been included in the acknowledgement. The correction to the book is available at https://doi.org/10.1007/978-981-19-9737-2_6.

Contents

Chapter 1
Background

Heavy rainfall in extreme climates often causes natural disasters such as floods, landslides, and debris flows. Rainfall-induced slope instabilities are major geological natural disasters (Glade 1998; Dai et al. 1999; Iverson 2000; Lee and Pradhan 2007; Li et al. 2016a, b; Wu et al. 2020) that can result in considerable loss of life and damage to infrastructure. Extreme events such as storms, which are becoming more severe because of climate change, can trigger fatal landslides. Storm-induced slope failures frequently occur because of rainfall infiltration, particularly in tropical areas (Fourie 1996; Cevasco et al. 2014). Global climate change in many mountainous areas could lead to more severe fluctuations in rainfall, and trigger of soil slope deformations and even slope instability because of the alteration of intensity, frequency, and quantity of rainfall (Dixon and Brook 2007; Jeong et al. 2008). The influence of climate change on rainfall characteristics has the potential to alter the stability of unsaturated soil slopes. Rainfall infiltration causes a decrease in matric suction and an increase in moisture content and hydraulic conductivity in unsaturated soils. The rainfall intensity and duration, initial water table, and hydraulic conductivity are the parameters that significantly affect slope stability (Ng and Shi 1998). An increase in pore-water pressure can reduce the effective stress and thereby weaken the shear strength of slopes. Complex geological environment and human engineering activities are also significant factors of slope instability under rainfall conditions.

Rainfall-induced slope failures have been examined based on experimental modeling, analytical and numerical methods (Ng and Shi 1998; Iverson 2000; Chen et al. 2005; Wu et al. 2009, 2016, 2020; Zhu et al. 2019). Laboratory and field experiments have been carried out to examine the infiltration mechanisms associated with rainfall-induced slope failures (e.g., Lee et al. 2011; Wu et al. 2015, 2017, 2018). Many numerical and analytical studies have investigated the hydraulic responses of slopes to rainfall infiltration and the stability of slopes under such conditions (Iverson 2000; Cai and Ugai 2004; Wiles 2006; Ali et al. 2014; Zhu et al. 2022). Numerical analysis solves for the matric suction or pressure head distribution in a soil slope with

© The Author(s) 2023
L. Wu and J. Zhou, *Rainfall Infiltration in Unsaturated Soil Slope Failure*,
SpringerBriefs in Applied Sciences and Technology,
https://doi.org/10.1007/978-981-19-9737-2_1

varying permeability, and considering different surface conditions of a soil. Many cases demonstrate that change in rainfall patterns may lead to slope failure due to infiltration. The results can provide an indication of the potential influence of climate change on shallow landslides in many mountainous areas (Kim et al. 2012).

Slope stability problems are commonly encountered in engineering projects. Many slope failures are attributed to water infiltration (Cho and Lee 2002; Cai and Ugai 2004). Matric suction is crucial to the stability of soil slopes because dissipation of matric suction leads to decrease in shear strength of unsaturated soils. Slope failure is closely related to the rainfall-induced transient infiltration of slopes in unsaturated soils (Fredlund and Rahardjo 1993; Wu et al. 2020). Shallow landslides are related to periods of intense rainfall. Engineering activities can result in severe geological and environmental issues. Slope failures induced by engineering activities may occur and may progress into landslides. The internal mechanics of slope movements are stress redistribution and the consequent changes in engineering–geological conditions (Marschalko et al. 2012).

Landslide forecasting should take into account rainfall infiltration into soil slopes. The models in predicting the timing and location of landslides are related to dynamic water infiltration in soil slope. The coupled process in an unsaturated soil is of major interest because of its implications for disaster prevention and environmental issues. Analytical approaches have been developed to provide a basic understanding of unsaturated infiltration in terms of the coupling effect (Wu et al. 2020). Meanwhile, numerical approaches provide a powerful tool for solving complex, nonlinear infiltration into unsaturated soils. These numerical models effectively investigate the coupled hydro-mechanical problem involved in unsaturated rooted slope stability issues (Oka et al. 2010; Wu et al. 2020; Zhu et al. 2022).

1.1 Rainfall Infiltration Equation

According to the modified Green–Ampt model (Mein and Larson 1973), the water infiltration process in soils during uniform precipitation can be divided into two stages: a stage controlled by rainfall intensity and a stage controlled by pressure head. The infiltration rate (f_a) determined by rainfall intensity (q_r) can be written as:

$$f_a = q_r \cos \beta \tag{1.1}$$

where q_r is the rainfall intensity; and β is the slope angle.

The infiltration rate (f_b) determined by the pressure head can be expressed as:

$$f_b = k_s \frac{z_f \cos \beta + s_f}{z_f} \tag{1.2}$$

where k_s represents the permeability coefficient at saturation; s_f represents the suction head of the wetting front; and z_f represents the wetting front depth.

If $f_a = f_b$, the water ponding time (t_p) can be obtained as:

$$t_p = \frac{s_f(\theta_s - \theta_i)}{\cos^2 \beta (q_r/k_s - 1)q_r}$$ (1.3)

where θ_s and θ_i are the saturated moisture content and initial moisture content, respectively.

According to the water balance principle and Darcy law, wetting front movement during rainfall can be calculated as follows:

$$\begin{cases} z_f = [q_r \cos \beta]\frac{t}{(\theta_s - \theta_i)}, & t < t_p \\ \frac{dz_f}{dt} = \frac{1}{(\theta_s - \theta_i)}\left[k_s \frac{z_f \cos \beta + s_f}{z_f}\right], & t \geq t_p \end{cases}$$ (1.4)

Equation (1.4) is the GA model suitable for slopes (Chen and Young 2006).

A penetration test of saturated sand layers was conducted and found a quantitative relationship between the infiltration velocity of water in the soil and the head loss, namely Darcy law:

$$q = -k_s \frac{dH}{dL}$$ (1.5)

where q is the flux (discharge per unit area, with units of length per time, m/s), H is the total water head, and L is the seepage length.

When soil mass is unsaturated, most scholars believe that Darcy law can also be used for analyzing water movement in unsaturated soils. Richards (1931) extended the Darcy law of saturated soil to the unsaturated infiltration and introduced the continuity equation to derive the equation of motion of the unsaturated soil–water flow, namely the Richards' equation. The permeability coefficient (k) is expressed as a function of matric suction, and Darcy's law can be expressed as:

$$q = k\nabla H$$ (1.6)

where ∇H is a hydraulic gradient, including two components of gravity and suction.

Continuity equation:

$$\frac{\partial \theta}{\partial t} = -\nabla \cdot q$$ (1.7)

Then:

$$\nabla \cdot [k\nabla H] = \frac{\partial \theta}{\partial t}$$ (1.8)

Substituting Eq. (1.6) into Eq. (1.8), one can obtain:

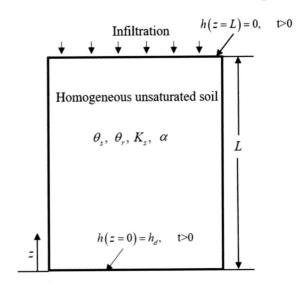

Fig. 1.1 1D rainfall infiltration model

$$\nabla \cdot [k(h)\nabla(h+z)] = \frac{\partial \theta}{\partial t} \tag{1.9}$$

As shown in Fig. 1.1, the Richards' equation governing one-dimensional vertical infiltration in unsaturated soils can be written as:

$$\nabla \cdot [k(h)\nabla h] + \frac{\partial k(h)}{\partial z} = \frac{\partial \theta}{\partial t} \tag{1.10}$$

1.2 Infiltration Equation for Unsaturated Slopes

The 2D generalized RE for unsaturated infiltration is expressed as (Ku et al. 2017; Wu et al. 2020):

$$\frac{\partial}{\partial x}\left[K_x(h)\frac{\partial H}{\partial x}\right] + \frac{\partial}{\partial z}\left[K_z(h)\frac{\partial H}{\partial z}\right] = C(h)\frac{\partial H}{\partial t} \tag{1.11}$$

where $K_z(h)$ and $K_x(h)$ are the permeability coefficients along the vertical direction and lateral direction in unsaturated soils, respectively. To study the groundwater flow of the unsaturated slope (Fig. 1.2), the RE needs to be rotated. Total head can be written as:

$$H = h + E \tag{1.12}$$

and elevation head E can be described as:

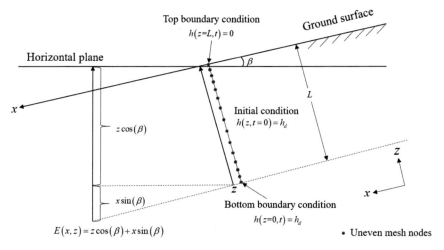

Fig. 1.2 A slope infiltration model

$$E = x \sin \beta + z \cos \beta \tag{1.13}$$

By substituting Eqs. (1.12) and (1.13) into Eq. (1.11), Eq. (1.11) can be re-expressed as:

$$\frac{\partial}{\partial x}\left[K_x(h)\left(\frac{\partial h}{\partial x} - \sin \beta\right)\right] + \frac{\partial}{\partial z}\left[K_z(h)\left(\frac{\partial h}{\partial z} + \cos \beta\right)\right] = C(h)\frac{\partial h}{\partial t} \tag{1.14}$$

According to Iverson's model (2000), the modified RE only considers infiltration in the vertical direction, which can be given by:

$$\frac{\partial}{\partial z}\left[K_z(h)\left(\frac{\partial h}{\partial z} + \cos \beta\right)\right] = \frac{\partial \theta}{\partial t} \tag{1.15}$$

1.3 Linearized Richards' Equation

Combined with an exponential model, Eq. (1.15) is linearized. Here, a new parameter h^* is defined as:

$$h^* = e^{\alpha h} - \lambda \tag{1.16}$$

where λ is the key parameter for solving the linearized Richards' equation in the conversion method, which can be defined as a constant $\lambda = e^{\alpha h_d}$ (Tracy 2006; Liu et al. 2015); h_d is the pressure head value when the soil is dry. The relative permeability coefficient is expressed as:

$$K(h) = \frac{K_z(h)}{k_s} = e^{\alpha h} \tag{1.17}$$

Taking the derivative of Eq. (1.16) with respect to z, one can obtain:

$$\frac{\partial h^*}{\partial z} = \alpha e^{\alpha h} \frac{\partial h}{\partial z} \tag{1.18}$$

Equation (1.18) can be re-stated as follows:

$$\frac{\partial h}{\partial z} = \frac{1}{\alpha} e^{-\alpha h} \frac{\partial h^*}{\partial z} \tag{1.19}$$

Furthermore, substituting Eq. (1.16) into Eq. (1.19), one can obtain:

$$K \frac{\partial h}{\partial z} = e^{\alpha h} \left(\frac{1}{\alpha} e^{-\alpha h} \frac{\partial h^*}{\partial z} \right) \tag{1.20}$$

Equation (1.20) can be rewritten as:

$$K \frac{\partial h}{\partial z} = \frac{1}{\alpha} \frac{\partial h^*}{\partial z} \tag{1.21}$$

Equation (1.17) can be derived from z again:

$$\frac{\partial K}{\partial z} = \alpha e^{\alpha h} \frac{\partial h}{\partial z} \tag{1.22}$$

Substituting Eq. (1.19) into Eq. (1.22), one can have:

$$\frac{\partial K}{\partial z} = \frac{\partial h^*}{\partial z} \tag{1.23}$$

An exponential model is employed to describe soil moisture (Gardner 1958):

$$\theta(h) = \theta_r + (\theta_s - \theta_r) e^{\alpha h} \tag{1.24}$$

where $\theta(h)$ is volumetric water content; θ_r is the residual volumetric water content.
The derivation of both sides of Eq. (1.24) with respect to time t has the following relationship:

$$\frac{\partial \theta}{\partial t} = (\theta_s - \theta_r) \frac{\partial K}{\partial t} = (\theta_s - \theta_r) \frac{\partial h^*}{\partial t} \tag{1.25}$$

Substituting Eqs. (1.19), (1.23), and (1.24) into Eq. (1.10), the linearized Richards' equation can be obtained:

$$\frac{\partial^2 h^*}{\partial z^2} + \alpha \frac{\partial h^*}{\partial z} = c \frac{\partial h^*}{\partial t} \tag{1.26}$$

where $c = \alpha(\theta_s - \theta_r)/k_s$.

Equation (1.26) is also expressed as:

$$K_a \frac{\partial^2 h^*}{\partial z^2} + K_\theta \frac{\partial h^*}{\partial z} = \frac{\partial h^*}{\partial t} \tag{1.27}$$

where $K_\theta = k_s/(\theta_s - \theta_r)$, $K_a = K_\theta/\alpha$.

The finite difference format of the linearized RE (Eq. 1.26) can be expressed as:

$$K_a \frac{h_{i+1}^{*n} - 2h_i^{*n} + h_{i-1}^{*n}}{\Delta z^2} + K_\theta \cos\beta \frac{h_{i+1}^{*n} - h_{i-1}^{*n}}{2\Delta z} = \frac{h_i^{*n} - h_i^{*n-1}}{\Delta t} \tag{1.28}$$

where i, Δz, Δt, and n denote the nodal point, grid size, time step, and time level, respectively.

It can be seen from Eqs. (1.26) and (1.27) that the nonlinear partial differential equation (Eq. 1.26) has been transformed into a linear partial differential equation. Once the linear partial differential equation is solved to obtain a numerical solution, the actual pressure head can be written as:

$$h(z, t) = \frac{1}{\alpha} \ln\left(h^*(z, t) + \lambda\right) \tag{1.29}$$

1.4 Unsaturated Soil Slope Stability Under Rainfall

Shear strength is a fundamental material property that is required to address a variety of engineering problems including bearing capacity, slope stability, lateral earth pressure, pavement design, and foundation design. Recently, many researches have focused on the shear strength of unsaturated soils (Fredlund et al. 1996; Lu et al. 2010).

According to Mohr–Coulomb criterion and effective stress, the shear strength of saturated soils can be expressed as:

$$\tau_f = c' + \sigma' \tan\varphi' \tag{1.30}$$

where τ_f is the shear strength (kPa), c is the cohesion (kPa), σ_n is the normal stress acting on the failure surface (kPa), and φ is the angle of internal friction (°). Cohesion and cohesive shear strength are due to chemical bonding between soil particles and surface tension within the water films (Lu and Likos 2006). Frictional shear strength ($\sigma_n \tan \varphi$) is owing to internal friction between soil particles that depends on the normal stress acting on the failure surface.

Engineering practices indicate that the shear strength equation of saturated soils can meet the engineering requirements. The shear strength parameters are also influenced by matric suction. With an increase in matric suction, c and φ increase, which depends on soil texture and structure. Soil shear strength significantly increases with an increase in net normal stress, matric suction, and the parameters of shear strengths.

However, several phases of unsaturated soils make the shear strength equation of saturated soils difficult to apply. Therefore, some studies on the shear strength criteria of unsaturated soils have been carried out. There are main representative shear strength criteria here.

Bishop (1959) developed a shear strength criterion for unsaturated soils:

$$\tau_f = c' + \left[(\sigma - u_a)_f + \chi(u_a - u_w)_f\right] \tan \varphi' \tag{1.31}$$

where τ_f is the shear strength of unsaturated soils; c' and φ' are the effective cohesion and friction angle, respectively; $(u_a - u_w)$ is the matric suction; u_a is the pore air pressure; u_w is the pore-water pressure ($h = u_w/\gamma_w, \gamma_w = \rho_w g$); and χ is the function of the degree of saturation.

Based on two stress state variables, the following equation was developed to describe shear strength (Fredlund and Rahardjo 1993):

$$\tau_f = c' + (\sigma - u_a)_f \tan \varphi' + (u_a - u_w)_f \tan \varphi^b \tag{1.32}$$

where φ^b is the internal friction angle due to the distribution of matric suction.

Lu and Likos (2004) proposed a unified form of shear strength equation:

$$\begin{aligned} \tau_f &= c' + \chi_f(\sigma - u_a)_f \tan \varphi' + \chi_f(u_a - u_w)_f \tan \varphi' \\ &= c' + c'' + (\sigma - u_a)_f \tan \varphi' \end{aligned} \tag{1.33}$$

in which

$$c'' = \chi_f(u_a - u_w)_f \tan \varphi' \tag{1.34}$$

The first two terms in Eq. (1.33), c' and c'', represent shear strength due to the so-called apparent cohesion in unsaturated soils. In an unsaturated soil, the third term represents frictional shearing resistance provided by the effective normal force at the grain contacts. The apparent cohesion captured by the first two terms includes the classical cohesion c' representing shearing resistance arising from interparticle physicochemical forces, and the second term c'' describing shearing resistance arising from capillarity effects. The term c'' is defined as capillary cohesion hereafter. Physically, capillary cohesion describes the mobilization of suction stress $\chi(u_a - u_w)$ in terms of shearing resistance. The relationship between capillary cohesion and the maximum suction stress at failure, $\chi_f(u_a - u_w)_f$ is defined as shear strength also affects the water movement of the soils (Eudoxie et al. 2012).

Slope failure in unsaturated soil regions induced by rainfall is due to shear strength of unsaturated soils (Fredlund and Rahardjo 1993; Lu and Likos 2004; Guzzetti et al. 2008; Muntohar and Liao 2009). Both rainfall characteristics (rainfall intensity and duration) and soil permeability may influence failure mechanism.

The soil slope stability was commonly followed by stability analysis according to the pressure head and/or the stress condition within the soil slope profile. Various techniques were employed to compute factor of safety (F_s), and the conventional limit equilibrium methods (Alonso et al. 2010). The limit equilibrium approach is mostly effective for slope failure with a small depth compared with their length and breadth. A slope sliding at a depth happens as the driving stress contributing to failure exceeds the anti-slip stress offered by the soil mass strength. Namely, sliding can occur at a particular depth as follows:

$$F_s(z, t) = \frac{\tan \varphi'}{\tan \beta} + \frac{c' - h(z, t)\gamma_w \tan \varphi^b}{W \sin \beta \cos \beta} \tag{1.35}$$

where $F_s(z, t)$ is the safety factor over depth and time; W is the weight of the sliding mass; and γ_w represents the unit weights of water.

Equation (1.35) can be re-arranged as:

$$h(z, t) = \frac{c'}{\gamma_w \tan \varphi^b} - \frac{\gamma_{sat} z \sin \beta \cos \beta}{\gamma_w \tan \varphi^b} \left(F_s - \frac{\tan \varphi'}{\tan \beta} \right) \tag{1.36}$$

in which, γ_{sat} represents the unit weight of the saturated soil. When F_s approaches 1, the infinite soil slope reaches a limit state. Based on Eq. (1.36), the limit-state pore-water pressure head can be obtained.

Rainfall-induced landslides may occur in unsaturated soils above the groundwater table, usually with shallow sliding surfaces parallel to the slope surface (Lu and Godt 2008), which involves 2D and 3D problems. However, an infinite slope model is usually used as a simplified model of the 2D or 3D issues with simple geometry and ignores the stress concentration, the practice sometime demonstrates its effectiveness for assessing shallow slope stability (Michalowski 2018).

Slope instabilities are often hydrologically initiated by the advancement of the wetting front alone (Muntohar and Liao 2010), a rise in groundwater level (Asch et al.1999; Montgomery et al. 2009), and positive pore-water pressure on the soil–bedrock boundaries (Baum et al. 2010). The most common mechanism for rainfall-induced landslides occurs when the soil slides on a low-conductivity layer. Rainfall infiltration leads to a rise in the pressure head, resulting in positive pore-water pressures (Iverson 2000; Muntohar and Liao 2010).

Generally, unsaturated soil slope failures happen most frequently during or after rain periods (Wu et al. 2020). The characteristics of water flow, change of pore-water pressure, and shear strength of soils are the major parameters related to slope stability analysis involving unsaturated soils that are directly affected by the boundary conditions (i.e., infiltration and evaporation) at the soil–atmosphere interface. The relative importance of soil properties, rainfall intensity, initial water table location, and slope geometry in inducing instability of soil slopes under different rainfall was investigated through a series of studies. Soil properties and rainfall intensity were found to be the primary factors controlling the slope instability due to rainfall, while the initial water table location and slope geometry only played a secondary role (Rahardjo et al. 2007).

The Green–Ampt model is a typical approximate infiltration model. Due to the simplicity and few parameters, the approximate infiltration model has become popular (Grimaldi et al. 2013). The classic GA model is only suitable for infiltration in horizontal soils. Therefore, modified GA models have been developed to describe the water infiltration in layered soils and slopes (e.g., Mein and Larson 1973; Chen and Young 2006; Kale and Sahoo 2011). Some modified infiltration models that account for rainwater redistribution have also been proposed (e.g., Corradini et al. 1997; Dou et al. 2014). These infiltration models have been extended to regional rainfall-runoff models for the hydrological prediction of catchments (Yuan et al. 2019). However, the actual infiltration process is very complicated and affected by many factors such as soil heterogeneity and rainfall conditions, and becomes difficult to be described accurately based on theoretical formulations (Srivastava et al. 2020). These theoretical equations generally tend to overestimate the factor of safety of soil slopes, resulting in slides and geological hazards (Kim et al. 2012). Some intelligent methods have been developed to predict the water infiltration into soils using machine learning techniques (Sihag et al. 2018).

Hydrological responses and slope factor of safety due to rainfall are concerned from a perspective of hydro-mechanical coupling. Coupled and uncoupled hydro-mechanical behaviors in unsaturated soils have been carried out to characterize the physical responses of unsaturated infiltration (i.e., variation of soil moisture, matric suction, effective stress, shear strength, and slope stability) (Casini 2013). The coupled issues are strongly linked in unsaturated soil slopes due to water infiltration, and the coupled poromechanical model actually examines the behavior and stability of rooted soils subjected to rainfall (Kim et al. 2012). Pressure heads generated in the uncoupled analysis are employed to examine deformation or soil slope stability (Cai and Ugai 2004; Yoo and Jung 2006). The accuracy and computational efficiency of the uncoupled analysis highly depend on the selected time increments (Huang and Lo 2013). The soil hydraulic and mechanical responses are calculated simultaneously in the coupled analysis. The coupled analysis produced a reasonably well defined wetting front and a lower critical F_s for unsaturated soil slopes. The coupled investigation could produce more accurate assessment of soil slope stability due to water infiltration and demonstrate a better physical representation of water infiltration and stress variation within unsaturated soil slopes.

More and more methods highlight the role of vegetation because of their interception role of the canopy and the root characteristics. Meanwhile, recent studies indicate that vegetation cannot control the rainfall-induced shallow landslide distribution (Emadi-Tafti et al. 2021). Some researches focus on the effect of roots on root–soil composite strength, or saturated hydraulic conductivity (Alessio 2019). The more complex the root architecture is, the stronger the root-composite strength becomes, while the faster the rainfall infiltrates. It has generally been concluded that vegetation roots mechanically and hydrologically affect slope stability. The plant roots seem act as a positive function in root-composite strength, while a negative role in water infiltration. Plant roots have various architectures in different land ecosystems and climatic conditions (Ma et al. 2018). Increasing studies related to soil–root complex focus on the root architectures (Burylo et al. 2011; Li et al. 2016a, b). One major controversy exists, e.g., the plant roots play positive role and enhance slope strength (Arnone et al. 2016). The roots could advance rainfall infiltration, thus contributing an adverse effect on slope stability (Ghestem et al. 2011; Garg et al. 2015). The root–soil composite strength and the hydraulic conductivity are of utmost importance for the rooted soil slope stability.

References

Alessio P (2019) Spatial variability of saturated hydraulic conductivity and measurement based intensity-duration thresholds for slope stability, Santa Ynez Valley, CA. Geomorphology 342:103–116

Ali A, Huang JS, Lyamin AV, Sloan SW, Cassidy MJ (2014) Boundary effects of rainfall-induced landslides. Comput Geotech 61:341–354

Alonso EE, Pereira JM, Vaunat J, Olivella S (2010) A microstructurally based effective stress for unsaturated soils. Géotechnique 60(12):913–925

Arnone E, Caracciolo D, Noto LV, Preti F, Bras RL (2016) Modeling the hydrological and mechanical effect of roots on shallow landslides. Water Resour Res 52(11):8590–8612

Asch TWJV, Buma J, Beek LPHV (1999) A view on some hydrological triggering systems in landslides. Geomorphology 30(1):25–32

Baum RL, Godt JW, Savage WZ (2010) Estimating the timing and location of shallow rainfall-induced landslides using a model for transient, unsaturated infiltration. J Geophys Res Earth Surf 115(F3):1–26

Bishop AW (1959) The principle of effective stress. Tekn Ukebl 39:859–863

Burylo M, Hudek C, Rey F (2011) Soil reinforcement by the roots of six dominant species on eroded mountainous marly slopes (Southern Alps, France). Catena 84(1):70–78

Cai F, Ugai K (2004) Numerical analysis of rainfall effects on slope stability. Int J Geomech 4(2):69–78

Casini F (2013) Coupled processes during rainfall an experimental investigation on a silty sand. In: Poromechanics. ASCE, pp 1542–1549

Cevasco A, Pepe G, Brandolini P (2014) The influences of geological and land use settings on shallow landslides triggered by an intense rainfall event in a coastal terraced environment. Bull Eng Geol Environ 73:859–875

Chen L, Young MH (2006) Green–Ampt infiltration model for sloping surfaces. Water Resour Res 42(W07420):1–9

Chen CY, Chen TC, Yu WH, Lin SC (2005) Analysis of time-varying rainfall infiltration induced landslide. Environ Geol 48:466–479

Cho SE, Lee SR (2002) Evaluation of surficial stability for homogeneous slopes considering rainfall characteristics. J Geotech Geoenviron Eng 128(9):756–763

Corradini C, Melone F, Smith RE (1997) A unified model for infiltration and redistribution during complex rainfall patterns. J Hydrol 192(1–4):104–124

Dai FC, Lee CF, Wang SJ (1999) Analysis of rainstorm-induced slide-debris flows on natural terrain of Lantau Island, Hong Kong. Eng Geol 51:279–290

Dixon N, Brook E (2007) Impact of predicted climate change on landslide reactivation: case study on Mam Tor, UK. Landslides 4:137–147

Dou HQ, Han TC, Gong XN, Zhang J (2014) Probabilistic slope stability analysis considering the variability of hydraulic conductivity under rainfall infiltration-redistribution conditions. Eng Geol 183:1–13

Emadi-Tafti M, Ataie-Ashtiani B, Hosseini SM (2021) Integrated impacts of vegetation and soil type on slope stability: a case study of Kheyrud Forest, Iran. Ecol Model 446:109498

Eudoxie GD, Phillips D, Springer R (2012) Surface hardness as an indicator of soil strength of agricultural soils. Open J Soil Sci 2:341–346

Fourie AB (1996) Predicting rainfall-induced slope instability. Proc Inst Civ Eng Geotechn Eng 119(4):211–218

Fredlund DG, Rahardjo H (1993) Soil mechanics for unsaturated soil. Wiley, New York

Fredlund DG, Xing A, Fredlund MD, Barbour SL (1996) The relationship of the unsaturated soil shear to the soil-water characteristic curve. Can Geotech J 33(3):440–448

Gardner WR (1958) Some steady stale solutions of the unsaturated moisture low equation with applications to evaporation from a water table. Soil Sci 85(4):228–232

Garg A, Coo JL, Ng CWW (2015) Field study on influence of root characteristics on soil suction distribution in slopes vegetated with *Cynodon dactylon* and *Schefflera heptaphylla*. Earth Surf Process Landforms 40(12):1631–1643

Ghestem M, Sidle RC, Stokes A (2011) The influence of plant root systems on subsurface flow: implications for slope stability. BioScience 61(11):869–879

Ghestem M, Veylon G, Bernard A, Vanel Q, Stokes A (2014) Influence of plant root system morphology and architectural traits on soil shear resistance. Plant Soil 377(1–2):43–61

Glade T (1998) Establishing the frequency and magnitude of landslide-triggering rainstorm events in New Zealand. Environ Geol 35(2):160–174

Grimaldi S, Petroselli A, Romano N (2013) Curve-number/Green–Ampt mixed procedure for streamflow predictions in ungauged basins: parameter sensitivity analysis. Hydrol Process 27(8):1265–1275

Guzzetti F, Peruccacci S, Rossi M, Stark CP (2008) The rainfall intensity–duration control of shallow landslides and debris flows: an update. Landslides 5(1):3–17

Huang CC, Lo CL (2013) Simulation of subsurface flows associated with rainfall-induced shallow slope failures. J GeoEng 8(3):101–111

Iverson RM (2000) Landslide triggering by rain infiltration. Water Resour Res 36(7):1897–1910

Jeong S, Kim J, Lee K (2008) Effect of clay content on well-graded sands due to infiltration. Eng Geol 102:74–81

Kale RV, Sahoo B (2011) Green–Ampt infiltration models for varied field conditions: a revisit. Water Resour Manage 25(14):3505–3536

Kim J, Jeong S, Regueiro RA (2012) Instability of partially saturated soil slopes due to alteration of rainfall pattern. Eng Geol 147:28–36

Ku CY, Liu CY, Su Y, Xiao JE, Huang CC (2017) Transient modeling of regional rainfall-triggered shallow landslides. Environ Earth Sci 76(16):570

Lee S, Pradhan B (2007) Landslide hazard mapping at Selangor, Malaysia using frequency ratio and logistic regression models. Landslides 4(1):33–41

Lee LM, Kassim A, Gofar N (2011) Performances of two instrumented laboratory models for the study of rainfall infiltration into unsaturated soils. Eng Geol 117(1–2):78–89

Li WC, Dai FC, Wei YQ, Wang ML, Min H, Lee LM (2016a) Implication of subsurface flow on rainfall-induced landslide: a case study. Landslides 13(5):1109–1123

Li YP, Wang YQ, Ma C, Zhang HL, Wang YJ, Song SS, Zhu JQ (2016b) Influence of the spatial layout of plant roots on slope stability. Ecol Eng 91:477–486

Liu CY, Ku CY, Huang CC (2015) Numerical solutions for groundwater flow in unsaturated layered soil with extreme physical property contrasts. Int J Nonlinear Sci Numer Simul 16(7):325–335

Lu N, Godt J (2008) Infinite slope stability under steady unsaturated seepage conditions. Water Resour Res 44(11):W11404

Lu N, Likos WJ (2006) Suction stress characteristic curve for unsaturated soil. J Geotech Geoenviron Eng 132(2):131–142

Lu N, Godt JW, Wu DT (2010) A closed-form equation for effective stress in unsaturated soil. Water Resour Res 46(5):W05515

Lu N, Likos WJ (2004) Unsaturated soil mechanics. John Wiley and Sons, Inc

Ma C, Deng JY, Wang R (2018) Analysis of the triggering conditions and erosion of a runoff-triggered debris flow in Miyun County, Beijing, China. Landslides 15:2475–2485

Marschalko M, Yilmaz I, Bednárik M, Kubečka K (2012) Influence of underground mining activities on the slope deformation genesis: Doubrava Vrchovec, Doubrava Ujala and Staric case studies from Czech Republic. Eng Geol 147–148:37–51

Mein RG, Larson CL (1973) Modeling infiltration during a steady rain. Water Resour Res 9(2):384–394

Michalowski RL (2018) Failure potential of infinite slopes in bonded soils with tensile strength cut-off. Can Geotech J 55(4):477–485

Muntohar AS, Liao HJ (2009) Analysis of rainfall-induced infinite slope failure during typhoon using a hydrological–geotechnical model. Environ Geol 56:1145–1159

Muntohar AS, Liao HJ (2010) Rainfall infiltration: infinite slope model for landslides triggering by rainstorm. Nat Hazards 54:967–984

Montgomery DR, Schmidt KM, Dietrich WE, Mckean J (2009) Instrumental record of debris flow initiation during natural rainfall: Implications for modeling slope stability. J Geophys Res–Earth 114(F1):1–16

Ng CWW, Shi Q (1998) A numerical investigation of the stability of unsaturated soil slopes subjected to transient seepage. Comput Geotech 22:1–28

Oka F, Kimot S, Takada N (2010) A seepage-deformation coupled analysis of an unsaturated river embankment using a multiphase elasto-viscoplastic theory. Soils Found 50(4):483–494

Rahardjo H, Ong TH, Rezaur RB, Leong EC (2007) Factors controlling instability of homogeneous soil slopes under rainfall. J Geotech Geoenviron Eng 133(12):1532–1543

Richards LA (1931) Capillary conduction of liquids through porous mediums. Physics 1(5):318–333

Sihag P, Singh B, Sepah Vand A, Mehdipour V (2018) Modeling the infiltration process with soft computing techniques. ISH J Hydraul Eng 26(2):138–152

Srivastava A, Kumari N, Maza M (2020) Hydrological response to agricultural land use heterogeneity using variable infiltration capacity model. Water Resour Manage 34:3779–3794

Tracy FT (2006) Clean two and three-dimensional analytical solutions of Richards' equation for testing numerical solvers. Water Resour Res 42(8):8503, 1–11

Wiles TD (2006) Reliability of numerical modelling predictions. Int J Rock Mech Min Sci 43(3):454–472

Wu LZ, Zhang LM (2009) Analytical solution to 1D coupled water infiltration and deformation in unsaturated soils. Int J Numer Anal Met 33(6):773–790

Wu LZ, Huang RQ, Xu Q, Zhang LM, Li HL (2015) Analysis of physical testing of rainfall-induced soil slope failures. Environ Earth Sci 73(12):8519–8531

Wu LZ, Selvadurai APS, Zhang LM, Huang RQ, Huang JS (2016) Poro-mechanical coupling influences on potential for rainfall-induced shallow landslides in unsaturated soils. Adv Water Resour 98:114–121

Wu LZ, Zhou Y, Sun P, Shi JS, Liu GG, Bai LY (2017) Laboratory characterization of rainfall-induced loess slope failure. Catena 150:1–8

Wu LZ, Zhang LM, Zhou Y, Xu Q, Yu B, Liu GG, Bai LY (2018) Theoretical analysis and model test for rainfall-induced shallow landslides in the red-bed area of Sichuan. Bull Eng Geol Environ 77:1343–1353

Wu LZ, Huang RQ, Li X (2020) Hydro-mechanical analysis of rainfall-induced landslides. Springer

Yoo C, Jung HY (2006) Case history of geosynthetic reinforced segmental retaining wall failure. J Geotech Geoenviron Eng ASCE 132(12):1538–1548

Yuan W, Liu M, Wan F (2019) Calculation of critical rainfall for small-watershed flash floods based on the HEC-HMS hydrological model. Water Resour Manage 33:2555–2575

Zhu S R, Wu L Z, Shen Z H, et al (2019) An improved iteration method for the numerical solution of groundwater flow in unsaturated soils. Comput Geotech 114: 103113

Zhu SR, Wu LZ, Huang JS (2022) Application of an improved P(m)-SOR iteration method for flow in partially saturated soils. Comput Geosci 26(1):131–145

Chapter 2
Analytical Solution to Unsaturated Infiltration

2.1 Introduction

Rainfall infiltration in unsaturated soil slopes is a classic issue in geotechnical engineering (Conte and Troncone 2012; Iverson 2000; Morbidelli et al. 2018). Factors influencing the soil slope stability due to rainwater infiltration comprise the rainfall characteristics, the saturated permeability coefficient, the geometry of the slope, and the boundary and initial soil moisture conditions (Ali et al. 2014; Wu et al. 2020).

The spatial and temporal evolution of unsaturated infiltration involves a governing partial differential equation that is expressed by the Richards' equation (1931). The equation is highly nonlinear because the hydraulic conductivity and the pressure head depend on matric suction or moisture content. A number of exact and approximated analytical solutions to the infiltration equation were derived in past studies (e.g., Parlange et al. 1997; Basha 2011). Analytical solutions of the linearized infiltration equation were derived as an integral (Chen et al. 2003), as a Laplace transformation (Zhan et al. 2013), and as a Green's function (Basha 1999). While numerical approaches can effectively simulate complex nonlinear infiltrations into an unsaturated soil (Tracy 2006; Wu et al. 2020), analytical solutions can verify these numerical procedures. The incorporation of physically based infiltration expressions better quantifies the infiltration models and causes a more reliable prediction of the water infiltration (Basha 2011). Compared with numerical solutions, the analytical methods are widely used because they can accurately check numerical methods, and concisely represent the pressure head variations with rainfall infiltration (Godt et al. 2012; Wu et al. 2020). Several analytical solutions to Richards' equation were obtained using integral transform methods (Laplace and Fourier algorithms) and others (Qin et al. 2010). The interaction between soil infiltration and displacement was defined as hydro-mechanical coupling effect (Wu et al. 2020). The coupled equation is characterized by strong nonlinearity; thus, linearization is required when solving the equation (Li et al. 2013; Zhu et al. 2022). To derive the analytical solution to infiltration equation considering the coupled hydro-mechanical effect, the soil–water

© The Author(s) 2023
L. Wu and J. Zhou, *Rainfall Infiltration in Unsaturated Soil Slope Failure*,
SpringerBriefs in Applied Sciences and Technology,
https://doi.org/10.1007/978-981-19-9737-2_2

characteristic curve (SWCC) is represented using an exponential form, and then the analytical solution of pressure head with arbitrary initial conditions can be developed using integral transformation and other methods (Li and Wei. 2018; Wu et al. 2009, 2012, 2016, 2018, 2020).

Many natural slopes are covered with vegetation, which can hydraulically and mechanically affect the slope stability (Lynch 1995; Leung et al. 2017). Vegetation root water uptake changes the pore-water pressure or pressure head of slopes, which is defined as the hydraulic effect of vegetation on slopes (Leung et al. 2015; Ni et al. 2018; Wu et al. 2021). Water is transferred by roots from soil voids to upper stems and leaves. The root water uptake rate and hydraulic conductivities are influenced by a number of factors including the soil type, self-hydraulic resistance, and hydraulic resistance of surrounding soils (Nyambayo and Potts 2010). Based on exponential function SWCC, the analytical solutions for water unsaturated infiltration were developed using Green's function (Ng et al. 2015). The effect of hydrological conditions on the stability of vegetated soil slopes was investigated (Liu et al. 2016; Wu et al. 2022).

This chapter is to derive an analytical solution incorporating both the root contribution and hydro-mechanical coupling. The governing equation considering vegetated root and coupled infiltration and deformation is developed. The analytical solution is obtained using Green's function. Parametric analyses are carried out to investigate the effect of factors on the vegetated slope stability.

2.2 1D Analytical Solutions for Unsaturated Seepage

2.2.1 No Coupling

The one-dimensional infiltration equation can be expressed as (Richards 1931):

$$\frac{\partial}{\partial z}\left[K(h)\left(\frac{\partial h}{\partial z}+1\right)\right]=\frac{\partial \theta}{\partial t} \tag{2.1}$$

where h is the pressure head; z is the vertical direction; t is the rainfall time.

Based on Gardner model (1958) and variable definition ($h^* = e^{\alpha h} - \chi$ in Chap. 1, where α is the desaturation coefficient, $\chi = e^{\alpha h_d}$), the linearized Richards' equation can be obtained:

$$\frac{\partial^2 h^*}{\partial z^2}+\alpha\frac{\partial h^*}{\partial z}=c\frac{\partial h^*}{\partial t} \tag{2.2}$$

where $c = \alpha(\theta_s - \theta_r)/K_s$.

Tracy (2006) developed an analytical solution to transient infiltration in unsaturated soils:

$$h_a(z, t) = \frac{1}{\alpha} \ln\left(h_t^*(z, t) + h_s^*(z) + e^{\alpha h_d}\right) \tag{2.3}$$

in which,

$$h_s^*(z) = \left(1 - e^{\alpha h_d}\right) \times \left(1 - e^{-\alpha z}\right)/\left(1 - e^{-\alpha L}\right) \tag{2.4}$$

$$h_t^*(z, t) = \frac{2\left(1 - e^{\alpha h_d}\right)}{Lc} e^{\alpha(L-z)/2} \sum_{m=1}^{\infty} (-1)^m \left(\frac{\lambda_m}{\mu_m}\right) \sin(\lambda_m z) e^{-\mu_m t} \tag{2.5}$$

where $\lambda_m = m\pi/L$ and $\mu_m = \left(\alpha^2/4 + \lambda_m^2\right)/c$.
For unsaturated slopes (Fig. 2.1), the governing equations are modified as follows:

$$\frac{\partial}{\partial z}\left[K_z(h)\left(\frac{\partial h}{\partial z} + \cos \beta\right)\right] = \frac{\partial \theta}{\partial t} \tag{2.6}$$

where β is the slope angle. The analytical solution of the seepage equation along the z-axis can be expressed as:

$$h_t^*(z, t) = \frac{2\left(1 - e^{\alpha h_d}\right)}{Lc} e^{\alpha \cos \beta(L-z)/2} \sum_{m=1}^{\infty} (-1)^m \left(\frac{\lambda_m}{\mu_m}\right) \sin(\lambda_m z) e^{-\mu_m t} \tag{2.7}$$

$$h_s^*(z) = \left(1 - e^{\alpha h_d}\right) \frac{1 - e^{-\alpha \cos \beta z}}{1 - e^{-\alpha \cos \beta L}} \tag{2.8}$$

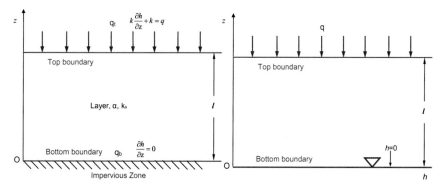

Fig. 2.1 Top and bottom flux boundaries for a soil profile with a finite thickness. The soil layer is between $z = 0$ and $z = l$, where z is the vertical coordinate: **a** impermeable boundary; and **b** groundwater level at the bottom

2.2.2 Hydro-mechanical Coupling

Based on Darcy's law, mass and momentum conservation, the equation that governs 1D hydro-mechanical coupling in unsaturated soils, including the consideration of increases in the water table, can be given by (Lloret 1987; Wu et al. 2020):

$$\frac{\partial}{\partial z}\left[k\frac{\partial}{\partial z}(h+z)\right] = \left(n\beta_w S_r \gamma_w + n\frac{\partial S_r}{\partial h}\right)\frac{\partial h}{\partial t} - S_r \alpha_c \frac{\partial \varepsilon_z}{\partial t} \tag{2.9}$$

$$\frac{\partial}{\partial z}\left[E\varepsilon_v - \frac{E}{F}(u_a - u_w)\right] + [nS_r \rho_w + (1-n)\rho_s]g = 0 \tag{2.10}$$

where ε_v ($\varepsilon_v = \varepsilon_z$ for one-dimensional problem) is the total volumetric strain of the soil mass, which is greater than zero during compression and less than zero during swelling; n is the percentage of voids; S_r is the saturation; $(u_a - u_w)$ is the matric suction; α_c is the hydro-mechanical coupling coefficient ($0 \leq \alpha_c \leq 1$), which is determined by the bulk modulus of the solid skeleton and the bulk modulus of the solid soil (Wu et al. 2020); E is the elastic modulus of the soil that takes account of alterations in the net normal stresses (Wu et al. 2020); F is the elastic modulus of the soil due to variations in matric suction, which is assumed to be a function of stress; β_w is the compressibility of the fluid; ρ_w is the density of water; ρ_s is the density of the soil phase; and g is the acceleration of gravity.

Equations (2.9) and (2.10) govern the coupling of 1D deformation and seepage in unsaturated soils and account for the groundwater table changes.

On the assumption that pore-air pressure in the soil mass is kept constant; the derivative of Eq. (2.10) with respect to t can be written as:

$$\frac{\partial \varepsilon_v}{\partial t} = -\frac{\gamma_w}{F}\frac{\partial h}{\partial t} \tag{2.11}$$

From Eqs. (2.9) and (2.11), the following equation can be derived:

$$\frac{\partial}{\partial z}\left[k\frac{\partial}{\partial z}(h+z)\right] = \left(\beta_w \frac{\partial \rho}{\partial h} + n\frac{dS_r}{dh}\right)\frac{\partial h}{\partial t} + \frac{\gamma_w S_r \alpha_c}{F}\frac{\partial h}{\partial t} \tag{2.12}$$

Based on assumption that the coefficient of permeability and the moisture content vary exponentially with the pore-water pressure head, that is, $k(h) = k_s e^{\alpha h}$ and $\theta(h) = \theta_s e^{\alpha h}$ ($h < 0, t > 0$). Here, k_s is the saturated hydraulic conductivity, θ_s is the volumetric moisture at saturation. Because $S_r = \theta(h)/\theta_s$ when the fluid compressibility is neglected ($\beta_w = 0$), Eq. (2.12) can be written as follows:

$$\frac{\partial}{\partial z}\left[k\frac{\partial}{\partial z}(h+z)\right] = Me^{\alpha h}\frac{\partial h}{\partial t} \quad (t > 0) \tag{2.13}$$

where $M = \theta_s \alpha + \gamma_w \alpha_c / F$.

With dimensionless variables $Z = \alpha z$ and $T = \alpha^2 k_s t/M$ introduced into a function of $W(Z, T) = e^{\alpha h} \cdot e^{Z/2 + T/4}$, Eq. (2.13) can be rewritten as:

$$\frac{\partial W}{\partial T} = \frac{\partial^2 W}{\partial Z^2} \tag{2.14}$$

The boundaries comprise a base and top one in Fig. 2.1. In the literature of analytical solutions, the base boundary was usually assumed to coincide with a stationary groundwater table and the pressure head was set zero (Wu et al. 2020). However, in this book, a zero flux is considered at the base boundary. The hydraulic boundary condition is given by:

$$h\bigg|_{z=0} = 0 \tag{2.15}$$

or

$$k\frac{\partial h}{\partial z} + k\bigg|_{z=0} = 0 \tag{2.16}$$

The top boundary in Fig. 2.1 is controlled by pressure head or rainfall intensity (q) at the ground surface, and it is written as:

$$h\bigg|_{z=l} = h_0 \tag{2.17}$$

or

$$k\frac{\partial h}{\partial z} + k\bigg|_{z=l} = q(t) \tag{2.18}$$

in which, l is the depth in the 1D unsaturated infiltration model.

Based on a Fourier integral transformation (Ozisik 1989), the exact solution to Eq. (2.14) can be derived considering different boundaries (Wu et al. 2020).

2.3 2D Analytical Solutions of Rainfall Infiltration in Unsaturated Soils

The 2D Richards' equation in mixed format is expressed as:

$$\frac{\partial}{\partial x}\left[K_x(h)\frac{\partial h}{\partial x}\right] + \frac{\partial}{\partial z}\left[K_z(h)\left(\frac{\partial h}{\partial z} + 1\right)\right] = \frac{\partial \theta}{\partial t} \tag{2.19}$$

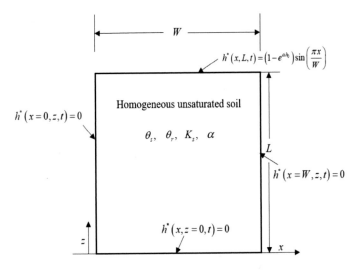

Fig. 2.2 Schematic diagram of 2D transient seepage model in unsaturated soils

in which, K_x and K_z are the hydraulic conductivities along x- and z-directions, respectively.

Similarly, the 2D linearized Richards' equation can be written as:

$$\frac{\partial^2 h^*}{\partial x^2} + \frac{\partial^2 h^*}{\partial z^2} + \alpha \frac{\partial h^*}{\partial z} = c \frac{\partial h^*}{\partial t} \tag{2.20}$$

The mathematical model is shown in Fig. 2.2. L and W represent the height and length, respectively. The normalized boundary conditions are as follows:

$$h(0, z, t) = h_{\mathrm{d}} \tag{2.21}$$

$$h(W, z, t) = h_{\mathrm{d}} \tag{2.22}$$

$$h(x, 0, t) = h_{\mathrm{d}} \tag{2.23}$$

$$h(x, L, t) = \ln\left(\left(1 - e^{\alpha h_{\mathrm{d}}}\right) \sin\left(\frac{\pi x}{W}\right) + e^{\alpha h_{\mathrm{d}}}\right) / \alpha \tag{2.24}$$

The normalized analytical solutions $h_{\mathrm{t}}^*(x, z, t)$ and $h_{\mathrm{s}}^*(x, z, t)$ for this two-dimensional model and can be expressed as follows (Tracy 2006):

$$h_t^*(x, z, t) = \frac{2(1 - e^{\alpha h_d})}{Lc} \sin\left(\frac{\pi x}{W}\right) e^{\alpha(L-z)/2} \sum_{m=1}^{\infty} (-1)^m \left(\frac{\lambda_m}{\mu_m}\right) \sin(\lambda_m z) e^{-\mu_m t}$$

$$(2.25)$$

$$h_s^*(x, z, t) = \left(1 - e^{\alpha h_d}\right) e^{\frac{\alpha(L-z)}{2}} \sin\left(\frac{\pi x}{W}\right) \frac{\sinh(\beta_1 z)}{\sinh(\beta_1 L)} \qquad (2.26)$$

where $\beta_1 = \sqrt{\alpha^2/4 + (\pi/W)^2}$.

2.4 Analytical Solution of Water Infiltration of Vegetated Slope Considering the Coupling Effects

Coupling between water infiltration and mechanical deformation in unsaturated soils is central to many natural and man-made systems in civil and environmental engineering. During water infiltration, the pore-water pressure or pressure head is redistributed, on one hand by the hydraulic properties of the unsaturated soils including retention characteristics and permeability, and, on the other hand, by the external loading due to climate conditions (rainfall intensity, duration, and evapotranspiration rate). Changes in the pore-water pressure or pressure head are generated by infiltration, which in turn modifies the hydraulic domain and induces deformations of the unsaturated soils. Alternatively, any variation in the mechanical loading can exert an effect on the infiltration process. It is indeed the hydro-mechanical coupled response of an unsaturated soil that is responsible for the most common instabilities associated with water infiltration: landslides and settlements, due to collapse or shear strength reduction (Thorel et al. 2011; Wu et al. 2020).

Recently, ecological protection technologies have become popular for slope environmental restoration and treatment (Tan et al. 2019; Broda et al. 2020). The stability analysis of vegetated slopes is a hot point in geotechnical engineering. Several studies have analyzed the stability of infinite vegetated slopes (Feng et al. 2020). However, few studies have been reported on the analytical solution of vegetated slope stability considering the hydro-mechanical coupling. Developing analytical solutions for water infiltration in unsaturated soil slopes is a significant issue in practical engineering. The main objective of this section is to derive an analytical solution considering both the root effect and hydro-mechanical behavior. The governing equation considering vegetated root and coupled infiltration-deformation is derived. The analytical solution is developed using Green's function method, which is then compared with the numerical solution. Parametric analyses are performed to investigate the effect of factors on the infinite vegetated slope stability.

2.4.1   Governing Equations of Rooted Unsaturated Soils

To simplify the problem, the following assumptions are made:

(1) The unsaturated slope is infinite, and the soil is isotropic and elastic. The groundwater level is assumed parallel to the slope surface, and the groundwater level is fixed.

(2) The growth direction of vegetation roots is perpendicular to the slope surface, and the root water uptake is simulated by adding a sink term (Raats 1979) to Richards equation.

(3) Soil skeleton is compressible, while water is incompressible.

An unsaturated soil slops with vegetation is shown in Fig. 2.3, L^* is the thickness of a soil (m), L_1^* is the depth of the rooted zone (m), L_2^* is the thickness of the unrooted zone (m). The Richards' equation in a reference coordinate system (oxz) can be expressed as:

$$\frac{\partial}{\partial x}\left[k(h)\frac{\partial h}{\partial x}\right] + \frac{\partial}{\partial z}\left[k(h)\frac{\partial h}{\partial z} + k(h)\right] = C(h)\frac{\partial h}{\partial t} \tag{2.27}$$

where h is the pressure head (m); $k(h)$ is the hydraulic conductivity (m/s); $C(h) = d\theta/dh$ is the differential water capacity (m^{-1}); θ is the volumetric water content; t is the infiltration duration (s). Equation (2.27) needs to be modified to be suitable for rooted soil slopes. The transformation relationship between the slope coordinate system (ox^*z^*) and the reference coordinate system (oxz) is described as follows:

$$x^* = x \cos \beta - z \sin \beta \tag{2.28a}$$

$$z^* = x \sin \beta + z \cos \beta \tag{2.28b}$$

Fig. 2.3 Graphical representation of rainfall-induced landslides

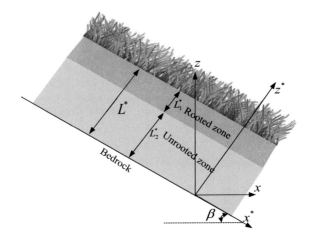

Substituting Eq. (2.28) into Eq. (2.27), according to Assumption (1), the improved Richards' equation for rainfall infiltration into soil slopes can be given by:

$$\frac{\partial}{\partial z^*}\left[k(h)\frac{\partial h}{\partial z^*}\right] + \frac{\partial k(h)}{\partial z^*}\cos\beta = C(h)\frac{\partial h}{\partial t} \qquad (2.29)$$

The sink term proposed by Raats (1979) is used to describe the water uptake of roots:

$$S(z^*) = g(z^*)T \qquad (2.30)$$

where $g(z^*)$ is the shape function of roots at depth z^* ($[m/s]^{-1}$); and T is the transpiration rate (m/s), which is affected by weather and leaf area index. Substituting Eq. (2.30) into Eq. (2.29), according to Assumption (2), the modified Richards equation considering the water uptake by vegetation roots can be obtained (Wu et al. 2022):

$$\frac{\partial}{\partial z^*}\left[k(h)\frac{\partial\psi}{\partial z^*}\right] + \frac{\partial k(h)}{\partial z^*}\cos\beta - S(z^*)\langle z^* - L_2^*\rangle = C(h)\frac{\partial h}{\partial t} \qquad (2.31)$$

where $\langle z^* - L_2^*\rangle = \begin{cases} z^* - L_2^*, & z^* \geq L_2^* \\ 0, & z^* < L_2^* \end{cases}$ is the sign function.

Equation (2.31) ignores the hydraulic coupling in unsaturated soils. According to the law of conservation of mass, the strict expression of the term $C(h)\frac{\partial h}{\partial t}$ on the right side of Eq. (2.31) is $\frac{1}{\rho}\frac{\partial}{\partial t}(\rho n S_r)$ (Kim 2000; Wu et al. 2020, 2022), where ρ, n, and S_r represent the density of water, soil porosity, and degree of saturation. $S_r = (\theta(h) - \theta_r)/(\theta_s - \theta_r)$, θ_s is the saturated water content, and θ_r is the residual water content. Then one can obtain:

$$\frac{\partial}{\partial t}(\rho n S_w) = \theta\frac{\partial\rho}{\partial t} + \rho\frac{\partial S_r}{\partial t}n + \rho S_r\frac{\partial n}{\partial t} \qquad (2.32)$$

the first term on the right side of Eq. (2.32) represents the water compression term, which is 0 according to Assumption (3); the second term is equivalent to $C(h)\frac{\partial h}{\partial t}$; the third term represents the soil skeleton term, and $\rho S_r\frac{\partial n}{\partial t} = \rho S_r\alpha_c\frac{\partial\varepsilon_v}{\partial t}$, where α_c is Biot's hydro-mechanical coupling parameter ($0 < \alpha_c \leq 1$), ε_v is volumetric strain. Therefore, Eq. (2.31) can be rewritten as:

$$\frac{\partial}{\partial z^*}\left[k(h)\frac{\partial h}{\partial z^*}\right] + \frac{\partial k(h)}{\partial z^*}\cos\beta - S(z^*)\langle z^* - L_2^*\rangle = C(h)\frac{\partial h}{\partial t} - S_r\alpha_c\frac{\partial\varepsilon_v}{\partial t} \qquad (2.33)$$

Substituting Eq. (2.11) into Eq. (2.33), the governing equation considering the hydro-mechanical coupling in unsaturated soil slopes can be obtained:

$$\frac{\partial}{\partial z^*}\left[k(h)\frac{\partial h}{\partial z^*}\right] + \frac{\partial k(h)}{\partial z^*}\cos\beta - S(z^*)\langle z^* - L_2^*\rangle = C(h)\frac{\partial h}{\partial t} + \frac{\gamma_w(\theta - \theta_r)\alpha_c}{(\theta_s - \theta_r)F}\frac{\partial h}{\partial t}$$

$$(2.34)$$

2.4.2 Analytical Solutions

According to Assumption (1), the bottom and surface boundary conditions of the slope are written as, respectively:

$$h(0, t) = 0 \quad t > 0 \tag{2.35}$$

$$\left[k(h)\frac{\partial h(L^*, t)}{\partial z^*} + k(h)\cos\beta\right] = q\cos\beta \quad t > 0 \tag{2.36}$$

where q is the rain intensity (m/s).

The hydraulic conductivity and the volumetric water content of unsaturated soils can be expressed as (Gardner 1958):

$$k(h) = k_s e^{\alpha h} \tag{2.37}$$

$$\theta(h) = \theta_r + (\theta_s - \theta_r)e^{\alpha h} \tag{2.38}$$

where k_s represents the saturated hydraulic conductivity (m/s); α is the desaturation coefficient (kPa^{-1}). Substituting Eq. (2.38) into Eq. (2.34), one has:

$$\frac{\partial}{\partial z^*}\left[k\frac{\partial h}{\partial z^*}\right] + \frac{\partial k}{\partial z^*}\cos\beta - S(z^*)\langle z^* - L_2^*\rangle = Me^{\alpha h}\frac{\partial h}{\partial t} \tag{2.39}$$

where $M = (\theta_s - \theta_r)\alpha + \gamma_w\alpha_c/F$.

The soil–water uptake function (Eq. 2.30) can be simplified as (Lynch 1995):

$$S(z^*) = T/L_1^* \tag{2.40}$$

Substituting Eqs. (2.37), (2.38), and (2.40) into Eq. (2.39) leads to:

$$\frac{\partial^2 k}{\partial z^{*2}} + \frac{\partial k}{\partial z^*}\alpha\cos\beta - T/L_1^*\langle z^* - L_2^*\rangle = \frac{M}{K_s}\frac{\partial k}{\partial t} \tag{2.41}$$

Here, the variables are defined as:

$$\begin{cases} Z = z^* \cos \beta \\ H = L^* \cos \beta \\ H_1 = L_1^* \cos \beta \\ H_2 = L_2^* \cos \beta \\ K = k/k_s \\ Q_a = q_a/k_s \\ Q_b = q_b/k_s \\ P = k_s/M \end{cases} \qquad (2.42)$$

where q_a and q_b represent the previous rain intensity (m/s) and current rain intensity (m/s).

Substituting Eq. (2.42) into Eq. (2.41) leads to:

$$\frac{\partial^2 K}{\partial Z^2} + \frac{\partial K}{\partial Z}\alpha - \frac{S(z^*/\cos\beta)(z^*/\cos\beta - L_2^*/\cos\beta)}{k_s \cos^2 \beta} = \frac{1}{P \cos^2 \beta}\frac{\partial K}{\partial t} \qquad (2.43)$$

By combining Eqs. (2.35), (2.36), and (2.43), the analytical solution of pressure head can be obtained using Green's function as follows:

$$K_{ste} = \begin{cases} \exp(-\alpha Z) + Q_a\left[\exp(-\alpha Z) - 1\right] \\ + \dfrac{T}{K_s \cos\beta H_1}\left[\exp(-\alpha Z) - 1\right](H - H_1), \qquad Z < H_2 \\ \\ \exp(-\alpha Z) + Q_a\left[\exp(-\alpha Z) - 1\right] + \dfrac{T}{K_s \cos\beta H_1} \\ \left\{ \begin{aligned} &\left[\exp(-\alpha Z) - 1\right](H - Z) + \exp(-\alpha Z) \\ &\left[Z - H_1 - \alpha^{-1}\exp(\alpha Z) + \alpha^{-1}\exp(\alpha H_1)\right] \end{aligned}\right\}, \qquad Z \geq H_2 \end{cases}$$

(Steady-state)   (2.44)

$$K = K_{ste} + 8\frac{\alpha P}{k_s}\cos^2\beta \exp\left[\frac{\alpha(H - Z)}{2}\right]$$
$$\times \sum_{i=1}^{\infty} \frac{\left[\lambda_i^2 + 0.25\alpha^2\right]\sin(\lambda_i H)\sin(\lambda_i Z)}{2\alpha + \alpha^2 H + 4H\lambda_i^2}G(t) \qquad \text{(Transient)} \qquad (2.45)$$

where

$$G(t) = \int_0^t (Q_a - Q_b)k_s \exp\left[-P\cos^2\beta(\lambda_i^2 + 0.25\alpha^2)(t - \tau)\right]d\tau \qquad (2.46)$$

λ_i is the ith root of the transcendental equation $\tan(\lambda H) + (2\lambda/a) = 0$. The transient solution changes with time. The "steady-state" infiltration evolves over formally infinite time, and thus the "steady-state" solution isn't related to time. However, transient infiltration depends on time, and the analytical solution to transient infiltration is called transient solution. Rainfall infiltration into soil slopes is commonly a transient issue, thus transient solutions are greatly meaningful.

When Q_b is a constant, Eq. (2.45) can be simplified as:

$$K = K_{ste1} - 8(Q_a - Q_b)\alpha \cos^2 \beta \exp\left[\frac{\alpha(H - Z)}{2}\right]$$

$$\times \sum_{i=1}^{\infty} \frac{\sin(\lambda_i H) \sin(\lambda_i Z) \exp\left[-P \cos^2 \beta\left(\lambda_i^2 + 0.25\alpha^2\right)t\right]}{2\alpha + \alpha^2 H + 4H\lambda_i^2} \qquad (2.47)$$

where

$$K_{ste1} = \begin{cases} \exp(-\alpha Z) + Q_b\left[\exp(-\alpha Z) - 1\right] \\ +\dfrac{T}{k_s \cos \beta H_1}\left[\exp(-\alpha Z) - 1\right](H - H_1), & Z < H_2 \\[4mm] \exp(-\alpha Z) + Q_b\left[\exp(-\alpha Z) - 1\right] + \dfrac{T}{k_s \cos \beta H_1} \cdot \\ \left\{\begin{bmatrix}\exp(-\alpha Z) - 1\end{bmatrix}(H - Z) + \exp(-\alpha Z) \\ \begin{bmatrix}Z - H_1 - \alpha^{-1}\exp(\alpha Z) + \alpha^{-1}\exp(\alpha H_1)\end{bmatrix}\right\}, & Z \geq H_2 \end{cases} \qquad (2.48)$$

The pore-water pressure head is calculated as follows:

$$h = \frac{\ln(K)}{\alpha} \qquad (2.49)$$

2.4.3 Comparison of Analytical and Numerical Solutions

The finite element method is employed to obtain the numerical solutions to Eqs. (2.34)–(2.36). In this book, COMSOL software is employed to implement numerical solutions, which has been used in geotechnical, hydraulic, and civil engineering.

The parameters adopted here are listed in Table 2.1, which includes the hydraulic parameters (k_s, α, θ_s, and θ_r) (Liu et al. 2016). The parameters describe the hydraulic properties of loess, and the parameters of vegetation roots are determined according to the statistics of shrubs (Feng et al. 2020).

The analytical and numerical solutions of pressure head at $t = 0$ h, 20 h, and 30 h are shown in Fig. 2.4. The analytical solutions at $t = 0$ h are obtained by Eq. (2.44), and those at $t = 20$ h and 30 h are obtained by Eq. (2.45). According to the root mean square error RMSE and coefficient of determination (R^2) shown in Fig. 2.4, it can be seen that RMSE between the numerical and analytical solutions is less than 0.005 m, and R^2 is very close to 1. That is, the error between the numerical and analytical solutions is very small. Compared with the numerical solution, the analytical solution concisely describes the pressure head variations with rainfall infiltration.

Table 2.1 Input parameters (Liu et al. 2016; Wu et al. 2022)

Parameters	Symbol	Value	Unit
Saturated hydraulic conductivity	k_s	1×10^{-6}	m/s
Desaturation coefficient	α	0.025	kPa^{-1}
Slope angle	β	30	°
Saturated water content	θ_s	0.47	–
Residual water content	θ_r	0.05	–
Suction-based modulus	F	10^3	kPa
Transpiration rate	T	5.21×10^{-8}	m/s
Rain intensity	q	5×10^{-7}	m/s
Soil thickness	L^*	5	M
Depth of the rooted zone	L_1^*	0.5	M

Fig. 2.4 Comparison of the analytical and numerical solutions

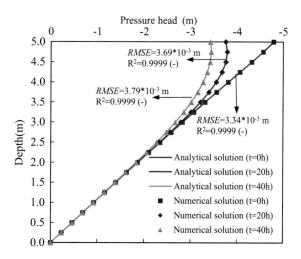

2.4.4 Parametric Analyses

Factor of safety (F_s) is an important indicator used to assess the stability of slopes, and slopes with F_s greater than 1 are generally considered to be stable, and vice versa. F_s is calculated based on the limit equilibrium method and strength theory of rooted unsaturated soils (Fredlund and Rahardjo 1993; Ku et al. 2018), as follows:

$$F_s = \frac{c + c_r}{(L^* - z^*)\gamma \cos \beta \sin \beta} + \frac{\tan \varphi}{\tan \beta} - \frac{\gamma_w h[(\theta - \theta_r)/(\theta_s - \theta_r)] \tan \varphi}{(L^* - z^*)\gamma \cos \beta \sin \beta} \qquad (2.50)$$

where c_r is the root cohesion (kPa), obtained by Wu–Waldron criterion (Wu 1976), i.e., $c_r = \varsigma T_s$, T_s is the root tensile strength (kPa), ς is the ratio of root cross-sectional area, which is 0.00025 in this book (Leung 2014). c and φ are the soil effective cohesion and effective frictional angle, while the rooted soil strength contributes to root cohesion. In this section, the effects of slope angle, rainfall intensity, transpiration rate, and suction-based modulus of elasticity on F_s and F_s ratio will be investigated. The F_s ratio, defined as the ratio of F_s during rainfall to the initial value ($t = 0$), shows the variations in F_s during rainfall and is also an important parameter for analyzing rainfall-include landslides.

2.4.4.1 Effect of Slope Angle

Here, the effects of slope angle on the pressure head, F_s and F_s ratio (defined as the ratio of F_s during rainfall to the initial value ($t = 0$)) are investigated. The parameters are as follows: Rain intensity is $0.7k_s$, the tensile strength of the roots is 10 MPa (this parameter is determined by the statistics of shrubs (Leung 2014)), the unit weight of the soil (γ) is 20 kN/m^3, the soil effective cohesion and effective frictional angle are 18 kPa and 28°, and other parameters are listed in Table 2.1.

Figure 2.5 represents the pressure head in slopes with slope angles of 10, 30, and 50° during rainfall. The conclusions can be drawn as follows (Fig. 2.5): (i) the pressure head of slopes decreases during rainfall; (ii) the pressure head in the shallow slope is more sensitive than that in the deep slope during rainfall; (iii) the smaller the slope angle, the larger the pressure head at the same position, which is mainly caused by the initial pressure head profile (the initial pressure head of the slope is negatively correlated with slope angle); (iv) the smaller the slope angle, the greater the variation in the pressure head at the same time, this is because the small slope angle has a positive effect on rainfall infiltration.

F_s (related to depth) with slope angles of 10, 30, and 50° during rainfall are described in Fig. 2.6a. Figure 2.6a states the following points: (a) F_s of the shallow slope is greater than that of the deep slope; (b) F_s decreases during rainfall, which can be explained in Fig. 2.5; (c) increase at $z = 4.5$ m due to the reinforcement effect of vegetation roots. F_s ratio F_s ratio with slope angles of 10, 30, and 50° during rainfall are represented in Fig. 2.6b. The larger the slope angle, the larger F_s ratio in Fig. 2.6b. This demonstrates that the larger the slope angle, the smaller F_s. In Fig. 2.6a, F_s suddenly the stability of a gentle slope is more sensitive to rainfall than that of the deep slope.

Fig. 2.5 Effect of slope angle on the pressure head

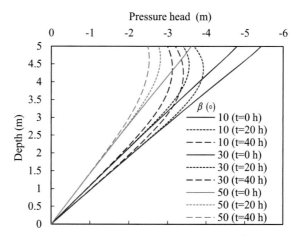

Fig. 2.6 Effect of slope angle on **a** F_s and **b** F_s ratio

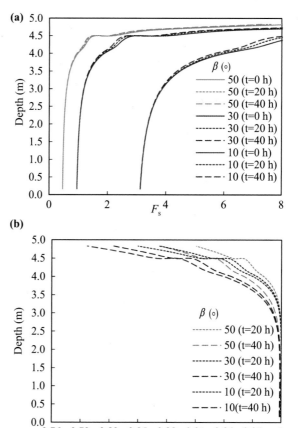

2.4.4.2 Effect of Rain Intensity

The variations in the dimensionless rainfall intensity over the duration of the rainfall event for both coupled and uncoupled analyses are shown in Fig. 2.7. The parameters are $k_s = 10^{-5}$ m/s, $\theta_s = 0.4$, $|F| = 5 \times 10^3$ kPa, and $\alpha = 0.01$ cm^{-1} (Van Genuchten 1980). The model height was 400 cm (Fig. 2.7). The value of F can be positive or negative. A negative F means an expansive soil where a suction decrease leads to soil volume increase, while a positive F denotes a collapsible soil where a soil suction decrease leads to soil volume decrease (Wu et al. 2020). Compared with the bottom boundary of a stationary groundwater table, the impermeable bottom boundary leads to more pronounced coupling effects in the lower part of the soil layer. It is also noted that the groundwater ponding occurs at the bottom boundary ($t = 15$ h, Fig. 2.7), particularly for an expansive soil ($F < 0$). Waterfall intensity plays a significant role in the advancement of the wetting front, and the pressure head profile moves more quickly as the rainfall intensity increases. The coupling effect is also closely linked with the rainfall intensity (Wu et al. 2020).

Here, the effect of rain intensity on the pressure head, F_s and F_s ratio are investigated. Figure 2.8 describes the pressure head of the slope with rainfall intensities of $0.5k_s$, $0.7k_s$ and $0.9k_s$. The remaining parameters are listed in Table 2.1. In Fig. 2.8, the smaller the rainfall intensity, the larger the pressure head at the same position.

Figure 2.9a, b describe F_s, and F_s ratio of the slope with rain intensities of $0.5k_s$, $0.7k_s$, and $0.9k_s$, respectively. The variation in F_s with time and depth is similar to that in Fig. 2.6a. The larger the rain intensity, the smaller F_s or F_s ratio of the slope. The stability of the shallow slope is more sensitive to that the deep slope under conductivity is an important parameter of soil seepage capacity, different rainfall intensities (Fig. 2.6). Saturated hydraulic conductivity is an important parameter of soil seepage capacity. Figure 2.10 represents the influence of saturated hydraulic

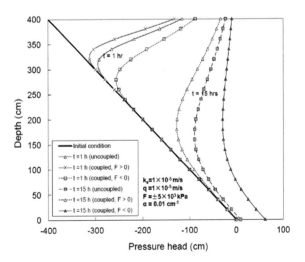

Fig. 2.7 Changes in the pressure head profile over time under coupled and uncoupled states

Fig. 2.8 Effect of rainfall intensity on the pressure head

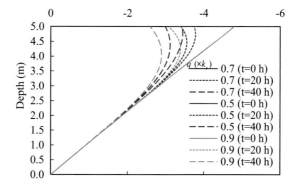

conductivity on F_s and F_s ratio of rooted soil slopes. With increasing saturated hydraulic conductivity, F_s and F_s ratio at the same depth decrease.

Fig. 2.9 Effect of rainfall intensity on the **a** F_s and **b** F_s ratio

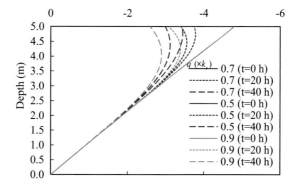

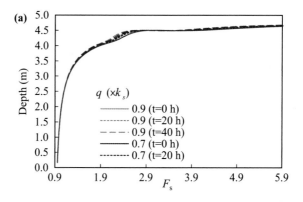

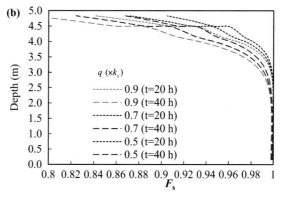

Fig. 2.10 Effect of saturated hydraulic conductivity on **a** F_s and **b** F_s ratio

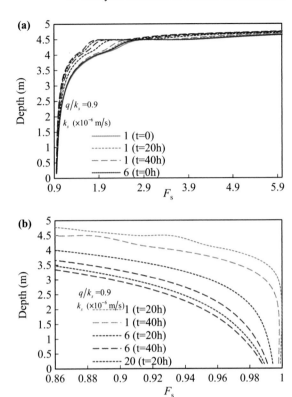

2.4.4.3 Effect of Transpiration Rate

The transpiration rate is an important parameter. The effect of transpiration rate on the pressure head, F_s and F_s ratio, is investigated here. The parameters are as follows: Rain intensity is $0.7k_s$, and other parameters are the same as those in Sect. 2.4.4.2.

Figure 2.11 represents the pressure head of unsaturated soil slopes with transpiration rates of 3, 4.5, and 6 mm/d. Transpiration rates depend on the vegetation and environmental conditions, and the transpiration rates are determined according to the statistics of shrubs (Leung and Ng 2013). The pressure head varies over time, and depth is the same as that in Sect. 2.4.4.1. The smaller the transpiration rate, the smaller the pressure head at the same position. The transpiration of vegetation reduces water content in slopes, causing an increase in soil suction. Figure 2.12a depicts F_s of slopes with transpiration rates of 3, 4.5, and 6 mm/d, respectively. The larger the transpiration rate, the larger safety factor of soil slopes. This is because the vegetation root uptake water reduces the water content of soil slopes. Figure 2.12b represents F_s ratio of slopes with transpiration rates of 3, 4.5, and 6 mm/d, respectively. The larger the transpiration rate, the smaller F_s ratio at the same position.

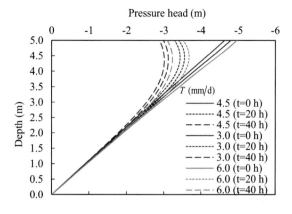

Fig. 2.11 Effect of transpiration rate on the pressure head

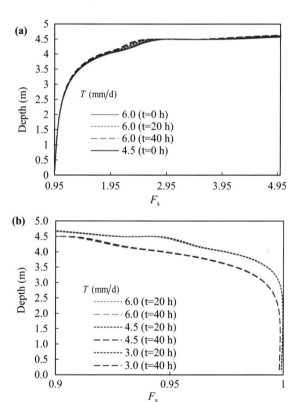

Fig. 2.12 Effect of transpiration intensity on **a** F_s and **b** F_s ratio

2.4.4.4 Effect of the Suction-Based Modulus of Elasticity

The suction-based modulus of elasticity (F) is key for hydro-mechanical coupling. The effect of suction-based modulus of elasticity on the pressure head, F_s, and F_s

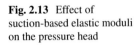

Fig. 2.13 Effect of suction-based elastic moduli on the pressure head

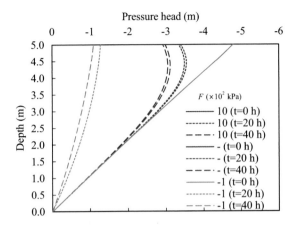

ratio, was investigated here. Three cases ($F = 10^3$ kPa, -10^2 kPa, and no hydro-mechanical coupling) are considered here. The governing equation without the hydro-mechanical coupling is Eq. (2.31), and the boundary conditions are the same as those considering the coupling effect. The transpiration rate in this section is 4.5 mm/d, and the other parameters are kept unchanged as Sect. 2.4.4.3.

Figure 2.13 represents the pressure head in soils with different suction-based moduli of elasticity. The variations of pressure head with suction-based modulus of elasticity are summarized as follows: (i) When F is positive, the hydro-mechanical coupling causes the increase of pore-water pressure head in the slope. Water flow in expansive soils is faster than that in collapsible soils. (ii) The effect of hydro-mechanical coupling on the pressure head in the slope becomes obvious with the decrease of the absolute value F.

Figure 2.14a represents F_s of soil slopes with different suction-based moduli of elasticity. Variation in safety factor with F is as follows: (i) When F is positive (negative), the hydro-mechanical coupling effect results in increases (decreases) of F_s. (ii) The effect of hydro-mechanical coupling on F_s becomes marked with decreasing absolute value of F.

Figure 2.14b describes F_s ratios of soil slopes with different suction-based moduli of elasticity. Figure 2.14b clearly demonstrates that small absolute value of F will significantly reduce the factor of safety of slopes.

2.5 Discussions and Conclusions

2.5.1 Discussions

F_s ratios caused by the hydro-mechanical coupling effect of bare slopes (transpiration rate is 0) and rooted slopes (transpiration rate is 4.5 mm/d) are compared. The

Fig. 2.14 Effect of
suction-based elastic moduli
on **a** F_s and **b** F_s ratio

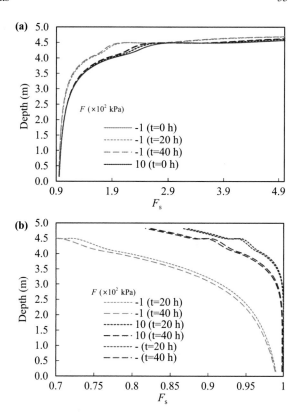

evaporation of bare soils caused by temperature is ignored. The F_s ratios of the bare
and rooted slopes at $t = 20$ h and 40 h are described in Fig. 2.15. The parameters
are the same as those in Sect. 2.4.4.4. In Fig. 2.15a, F_s ratios of the rooted slope
are smaller than those of the bare slope. The stability changes caused by the hydro-
mechanical coupling in rooted soil slopes are greater than those of bare slopes. A
similar conclusion can be drawn from Fig. 2.15b.

It should be emphasized that the proposed analytical solution has two main limitations: (i) This solution is only suitable for rooted slopes with uniform root architecture. Existing results (e.g., Ng et al. 2015; Liang et al. 2017) have indicated that
vegetation roots have complex shape, including exponential root, triangular root, and
uniform root. The book tries to obtain analytical solutions considering different root
architecture functions in future work. (ii) The boundary condition of soil slopes is
transformed from flow boundary to pressure head boundary due to saturated hydraulic
conductivity less than rain intensity, which may affect the results of analytical solutions. Although there still occur some limitations, the proposed analytical solution
incorporates the root effect and hydro-mechanical coupling for the first time. In
actual engineering, the proposed analytical solution can be easily applied to examine
rainfall-induced landslides in rooted soil regions.

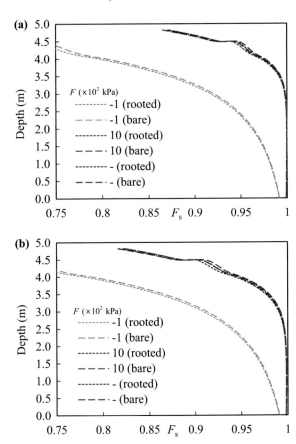

Figure 2.16 represents the variations in the dimensionless rainfall intensity over the duration of the rainfall event for both the coupled and uncoupled conditions. The parameters used are $k_s = 10^{-5}$ m/s, $\theta_s = 0.4$, $|F| = 5 \times 10^3$ kPa, and $\alpha = 0.01$ cm^{-1} (Van Genuchten 1980; Wu et al. 2020). The value of F can be positive or negative; a negative F means an expansive soil where a suction decrease leads to soil volume increase, while a positive F denotes a collapsible soil where a suction decrease leads to soil volume decrease (Wu et al. 2020). When $t = 1$ h, the wetting front moves to a depth of 120 cm in both the uncoupled analysis and coupled analysis with $F > 0$. However, with $F < 0$ the wetting front reaches 180 cm in the coupled analysis. When $t = 1$ h, the difference in the pressure head for the coupled ($F < 0$) and uncoupled conditions is 27.5 cm at the upper boundary, and the maximum difference in the pressure head within the unsaturated zone is 75.1 cm at a soil depth of 340 cm. When $t = 15$ h, the difference in the pressure head at the top boundary between the coupled ($F > 0$) and uncoupled conditions is 13.7 cm, and the maximum difference in the pressure head within the soil is 61.8 cm at a depth of 160 cm. The coupling effect becomes more apparent with increasing time, particularly in the bottom part

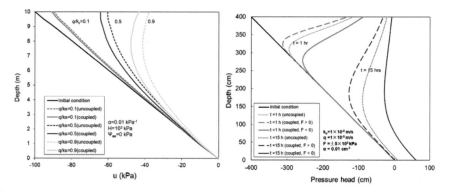

Fig. 2.16 Pore-water pressure profile for different boundaries

of the soil layer. Compared with the case where the base boundary coincides with the stationary groundwater table (Wu et al. 2020), the case with zero flux at the base boundary leads to more pronounced coupling effects in the lower part of the soil column. It is also noted that groundwater ponding occurs at the base boundary ($t =$ 15 h, Fig. 2.16).

As the infiltration time increases and the wetting front moves downward, the pressure head increases rapidly in the shallow depths of the unsaturated soils. Rainfall intensity plays an important role in the advancement of the wetting front, and the pressure head profile moves more quickly as the rainfall intensity increases. The coupling effect is also closely linked with the rainfall intensity. Other studies reported that the coupling effect is more noticeable in a shallow layer of an unsaturated soil (Wu et al. 2020). However, in this book, the coupling effect becomes more noticeable in the base layer, particularly with increasing time. As the rainfall accumulates at the base boundary, the coupled effect becomes more apparent there. Figure 2.16 describes the effect of boundary on the pressure profile considering the coupling effect. The boundary effect is more marked than the coupling. The zero-flux base boundary also leads to groundwater ponding in the lower part of the soils. The pressure head profile moves more quickly as the thickness of the soil layer decreases.

2.5.2 Conclusions

The governing equation considering hydro-mechanical coupling of unsaturated rooted slopes is derived. The analytical solution is obtained using Green's function method. This solution is compared with the finite element method. The parametric studies were carried out to investigate the effect of slope angle, rain intensity, suction-based elastic modulus, and transpiration rate on the pore-water pressure head and slope safety factor. The following conclusions can be obtained:

(1) We have developed an analytical solution that described transient groundwater flow in unsaturated soils using a Fourier integral transform, in which the groundwater level is allowed to advance, due to the zero flux boundary. The results indicate that the lower boundary condition plays a considerable role in the pressure head profile and that the coupling effect is more pronounced for a zero-flux boundary than a zero-pressure-head boundary. The coupling effect becomes more noticeable when ponding occurs.

(2) The proposed analytical solution is explicit and provides a basis for the numerical solution for rainfall infiltration into rooted soil slopes. The proposed analytical solution is simple in form and has few input parameters, which is a simple and effective method for analyzing rainfall infiltration in vegetated slopes.

(3) The larger the slope angle or rain intensity, the lower the pressure head, resulting in a smaller factor of safety for slopes. The larger the transpiration rate, the larger the pressure head, which causes an increase of safety factor of rooted soil slopes. The lower the absolute value of the suction-based elastic modulus, the more obvious the effect of hydro-mechanical coupling on the safety factor of rooted soil slopes.

References

Ali A, Huang JS, Lyamin AV, Sloan SW, Cassidy MJ (2014) Boundary effects of rainfall-induced landslides. Comput Geotech 61:341–354

Basha HA (1999) Multidimensional linearized nonsteady infiltration with prescribed boundary conditions at the soil surface. Water Resour Res 35(1):75–83

Basha HA (2011) Infiltration models for semi-infinite soil profile. Water Resour Res 47(8):192–198

Broda J, Franitza P, Herrmann U, Helbig R, Grosse A, Grzybowska-Pietras J, Rom M (2020) Reclamation of abandoned open mines with innovative meandrically arranged geotextiles. Geotext Geomembr 48(3):236–242

Chen JM, Tan YC, Chen CH (2003) Analytical solutions of one-dimensional infiltration before and after ponding. Hydrol Process 17(4):815–822

Conte E, Troncone A (2012) Stability analysis of infinite clayey slopes subjected to pore pressure changes. Géotechnique 62(1):87–91

Feng S, Liu HW, Ng CWW (2020) Analytical analysis of the mechanical and hydrological effects of vegetation on shallow slope stability. Comput Geotech 118:103335

Fredlund DG, Rahardjo H (1993) Soil mechanics for unsaturated soils. Wiley

Gardner WR (1958) Some steady-state solutions of the unsaturated moisture flow equation with application to evaporation from a water table. Soil Sci 85(4):228–232

Godt JW, Şener-Kaya B, Lu N, Baum RL (2012) Stability of infinite slopes under transient partially saturated seepage conditions. Water Resour Res 48(5):W05505

Iverson RM (2000) Landslide triggering by rain infiltration. Water Resour Res 36(7):1897–1910

Kim JM (2000) A fully coupled finite element analysis of water-table fluctuation and land deformation in partially saturated soils due to surface loading. Int J Numer Methods Eng 49(9):1101–1119

Ku CY, Liu CY, Su Y, Xiao J-E (2018) Modeling of transient flow in unsaturated geomaterials for rainfall-induced landslides using a novel spacetime collocation method. Geofluids 7892789

Leung T (2014) Native shrubs and trees as an integrated element in local slope upgrading. Ph.D. thesis, Department of Civil Engineering, University of Hong Kong

Leung AK, Ng CWW (2013) Analyses of groundwater flow and plant evapotranspiration in a vegetated soil slope. Can Geotech J 50(12):1204–1218

Leung AK, Garg A, Coo JL, Ng CWW, Hau BCH (2015) Effects of the roots of *Cynodon dactylon* and *Schefflera heptaphylla* on water infiltration rate and soil hydraulic conductivity. Hydrol Process 29(15):3342–3354

Leung AK, Boldrin D, Liang T, Wu ZY, Kamchoom V, Bengough AG (2017) Plant age effects on soil infiltration rate during early plant establishment. Géotechnique 68(7), 646–652

Lloret A, Gens A, Batlle F, Alonso E E (1987) Flow and deformation analysis of partially saturated soils. Proc., 9th European Conf. on Soil Mechanics, E. T. Hanrahan, T. L.L. Orr, and T. F. Widdis, eds., Balkema, Rotterdam, Netherlands, Dublin, 2:565–568

Liang T, Bengough AG, Knappett JA, MuirWood D, Loades KW, Hallett PD, Boldrin D, Leung AK, Meijer GJ (2017) Scaling of the reinforcement of soil slopes by living plants in a geotechnical centrifuge. Ecol Eng 109:207–227

Li J, Wei CF (2018) Explicit approximate analytical solutions of seepage-deformation in unsaturated soils. Int J Numer Anal Methods Geomech 42(7):943–956

Li JW, Wang HB, Zhang L (2013) Instability analysis of coupling seepage and stress field in unsaturated soil. Appl Mech Mater 446–447:1448–1455

Liu HW, Feng S, Ng CWW (2016) Analytical analysis of hydraulic effect of vegetation on shallow slope stability with different root architectures. Comput Geotech 80:115–120

Lynch J (1995) Root architecture and plant productivity. Plant Physiol 109(1):7–13. https://doi.org/10.1104/pp.109.1.7

Morbidelli R, Saltalippi C, Flammini A, Govindaraju RS (2018) Role of slope on infiltration: a review. J Hydrol 557:878–886

Ng CWW, Liu HW, Feng S (2015) Analytical solutions for calculating pore-water pressure in an infinite unsaturated slope with different root architectures. Can Geotech J 52(12):1981–1992

Ni JJ, Leung AK, Ng CWW (2018) Unsaturated hydraulic properties of vegetated soil under single and mixed planting conditions. Géotechnique 69(6):554–559

Nyambayo VP, Potts DM (2010) Numerical simulation of evapotranspiration using a root water uptake model. Comput Geotech 37(1–2):175–186

Ozisik MN (1989) Boundary value problems of heat conduction. Dover, New York

Parlange JY, Barry DA, Parlange MB et al (1997) New approximate analytical technique to solve Richards equation for arbitrary surface boundary conditions. Water Resour Res 33:903–906

Qin A, Sun DA, Tan Y (2010) Analytical solution to one-dimensional consolidation in unsaturated soils under loading varying exponentially with time. Comput Geotech 37(1–2):233–238

Raats PAC (1979) The distribution of the uptake of water by plants: inference from hydraulic and salinity data. In: Proceedings of the AGRIMED seminar on the movement of water and salts as function of the properties of soil under localized irrigation, Bologna, Italy, 6–9 Nov 1979

Richards LA (1931) Capillary conduction of liquids through porous mediums. Physics 1(5):318–333

Tan HM, Chen FM, Chen J, Gao YF (2019) Direct shear tests of shear strength of soils reinforced by geomats and plant roots. Geotext Geomembr 47(6):780–791

Thorel L, Ferber V, Caicedo B, Khokhar I (2011) Physical modelling of wetting-induced collapse in embankment base. Géotechnique 61(5):409–420

Tracy FT (2006) Clean two- and three-dimensional analytical solutions of Richards' equation for testing numerical solvers. Water Resour Res 42(8):W08503

Van Genuchten MT (1980) A closed form equation for predicting the hydraulic conductivity of unsaturated soils. Soil Sci Soc Am J 44(5):892–898

Wu TH (1976) Investigation of landslides on Prince of Wales Island. Ohio State University, Alaska

Wu LZ, Zhang LM (2009). Analytical solution to 1D coupled water infiltration and deformation in unsaturated soils. Int J Numer Anal Methods Geomech 33(6): 773–790

Wu LZ, Zhang LM, Huang RQ (2012) Analytical solution to 1D coupled water infiltration and deformation in two-layer unsaturated soils. Int J Numer Anal Methods Geomech 36(6): 798–816

Wu LZ, Zhang LM, Li X (2016) One-dimensional coupled infiltration and de-formation in unsaturated soils subjected to varying rainfall. Int J Geomech 16(2):06015004

Wu LZ, Xu Q, Wang T (2018) Incorporating hydromechanical coupling of un-saturated soils into the analysis of rainwater-induced groundwater ponding. Int J Geomech 18(6): 06018010

Wu LZ, Huang RQ, Li X (2020) Hydro-mechanical analysis of rainfall-induced landslides. Springer

Wu GL, Cui Z, Huang Z (2021) Contribution of root decay process on soil infiltration capacity and soil water replenishment of planted forestland in semi-arid regions. Geoderma 404:115289

Wu LZ, Cheng P, Zhou JT, Li SH (2022) Analytical solution of rainfall infiltration for vegetated slope in unsaturated soils considering hydro-mechanical effects. CATENA 206:105548

Zhan TLT, Jia GW, Chen YM, Fredlund DG, Li H (2013) An analytical solution for rainfall infiltration into an unsaturated infinite slope and its application to slope stability analysis. Int J Numer Anal Methods Geomech 37(12):1737–1760

Zhu SR, Wu LZ, Huang J (2022) Application of an improved P(m)-SOR iteration method for flow in partially saturated soils. Comput Geosci 26(1):131–145

Chapter 3
Numerical Solutions to Infiltration Equation

3.1 Introduction

Unsaturated infiltration issues occur in many fields, such as rainfall-induced soil slope failures (Wu et al. 2020a, b; Jiang et al. 2022), solute migration simulation (Cross et al. 2020), and coal seam water injection and coalbed methane extraction (Wang et al. 2020). Among them, the Richards' equation (Richards 1931) is the basic governing equation for the numerical simulation of unsaturated infiltration. The effective and reliable numerical solution of the Richards' equation is of great significance to scientific research and production in related fields. Before the advent of computers, the investigations on unsaturated infiltration mainly focused on the analytical solution of the infiltration equation, that is, the method of directly solving differential equations using relevant mathematical means. The studies on analytical solutions under certain conditions still have very important theoretical and practical significance. Analytical solutions to the Richards' equation can be obtained under some simplified conditions (Broadbridge et al. 2017). For example, Srivastava and Yeh (1991) employed an exponential model describing the soil-water character curve (SWCC) to derive transient analytical solutions for one-dimensional infiltration in homogeneous and layered soils. Parlange et al. (1999) solved a complex series solution of the one-dimensional Richards equation in order to obtain the infiltration flux. Tracy (2006) proposed two-dimensional and three-dimensional analytical solutions to rainfall infiltration into homogeneous soils based on exponential function. Wu et al. (2016, 2020a) used the Laplace transform method to obtain an analytical solution considering the coupling of infiltration and deformation in unsaturated soils during rainfall, and used it to analyze the stability of infinite unsaturated slopes due to rainfall infiltration. However, the permeability coefficients and soil–water characteristic curves are very complex in reality, which leads to highly nonlinear partial differential equations analytical solutions of which are very difficult to obtain. Therefore, numerical methods are often developed to solve the Richards' equation under

© The Author(s) 2023
L. Wu and J. Zhou, *Rainfall Infiltration in Unsaturated Soil Slope Failure*,
SpringerBriefs in Applied Sciences and Technology,
https://doi.org/10.1007/978-981-19-9737-2_3

common conditions (Zha et al. 2017; Ku et al. 2018; Zeng et al. 2018; Zhu et al. 2019; Eini et al. 2020).

With the development of computer techniques, numerical methods, namely finite difference method (FDM) (Liu et al. 2017), finite element method (FEM) (Crevoisier et al. 2009), and the finite volume method (FVM) (Liu 2017; Eymard et al. 2006), are increasingly used in infiltration analysis. The discretization of the time derivative is usually performed using the backward difference method (Pop and Schweizer 2011). For example, Patankar (1980) summarized FVM for numerical dis-cretization of heat transfer and fluid flow. Wang and Anderson (1982) introduced the application of finite difference and finite element methods for numerical simulation of groundwater infiltration and pollution propagation. Šimůnek et al. (2009) applied the finite element method to numerically solve the Richards equation and developed the commercial software Hydrus-1D. Zambra et al. (2011) constructed a finite volume method with high accuracy in space and time for solving nonlinear Richards equations. Ku et al. (2018) linearized the Richards' equation and developed a numerical solution to the unsteady groundwater infiltration issue by the collocation Trefftz method. Zeng proposed a modified Richards' equation and used FEM to more efficiently solve the variable saturated infiltration problem in heterogeneous soils. Chávez-Negrete et al. (2018) proposed an improved FDM combined with an adaptive step-size Crank–Nicolson method for solving the Richards' equation. Li et al. (2022) established the convergence selection criteria of grid size and time step in finite element simulation of unsaturated infiltration, which can better enhance the numerical accuracy. Of course, other advanced numerical methods have certain advantages in terms of computational accuracy, computational efficiency or ease of implementation under certain conditions when solving variable saturated infiltration issues. These methods include hybrid finite element methods (Bergamaschi and Putti 1999), cellular automata methods (Mendicino et al. 2006), FDM with non-orthogonal grids (An et al. 2010), finite analysis methods (Zhang et al. 2016), meshless methods (Herrera et al. 2009; Ku et al. 2021), and Chebyshev spectral methods (Wu et al. 2020b). Table 3.1 lists the development of some classical numerical methods for solving Richards' equation.

In this chapter, numerical solutions of water infiltration equation in unsaturated soils are examined. Some improved methods such as fractional-order and Chebyshev spectral methods will be applied to investigate nonlinear infiltration in homogeneous and layered soils. The results will be compared with software simulation.

3.2 Numerical Solutions

3.2.1 Finite Difference Method

The FDM to solve Richards' equation first divides the solution area into differential grids, replaces the continuous solution domain with a finite number of grid nodes,

Table 3.1 Some studies on the numerical solution of Richards' equation

Methodologies	Key improvements	References
Improved FEM (adaptive hp-FEM)	FEM combines adaptively finite elements of different spatial diameters (h) and polynomial degrees (p) to maximize the convergence rate of the iterative method	Solin and Kuraz (2011)
Improved FVM	Compared with FVM, improved method has better efficiency and accuracy	Liu (2017)
Generalized FDM	Compared with FDM, the proposed method has higher accuracy	Chávez-Negrete et al. (2018)
Linearization, Chebyshev spectral method	Compared with FDM, the proposed method has higher numerical accuracy	Wu et al. (2020b)
Linearization, modified meshless method	The accuracy and stability of the proposed functions are higher than those of the classical time stepping scheme	Ku et al. (2021)
Mixed FEM and the method of lines	The proposed method is easier to implement and is efficient and robust	Fahs et al. (2009)
FDM, modified Picard method combined with Anderson acceleration	Anderson acceleration significantly improves convergence speed and robustness of the modified Picard iteration	Lott et al. (2012)
FEM, Picard method, Picard/Newton method, L-scheme	A new scheme is proposed, the L-scheme/Newton method which is more robust and quadratically convergent	List and Radu (2016)
A linear domain decomposition method	The proposed scheme is more stable than the Newton scheme while remaining comparable in computational time	Seus et al. (2018)
Finite analytic method	This method can obtain more accurate numerical solutions and control the global mass balance better than modified Picard method	Zhang et al. (2016)
FEM, modified Picard method	The proposed modifications do not degrade the simulated results, while they cause more robust convergence performances	Zha et al. (2017)

(continued)

Table 3.1 (continued)

Methodologies	Key improvements	References
Modified L-scheme	The modified scheme is at least as fast as the modified Picard scheme, faster than the L-scheme, and is more stable than the Newton and the Picard scheme	Mitra and Pop (2019)

and stores the variables to be solved (pressure head, water content, etc.) in each grid node. The differential term in the Richards' equation is replaced by the corresponding difference quotient, so that the partial differential equation is converted into an algebraic difference equation, and a differential equation system containing a finite number of unknown variables at discrete points is obtained, that is, a linear equation system. The solution of the linear equation system is obtained, and the numerical solution of the variables on the grid nodes is obtained. The principle is simple and easy to implement, and it is widely used. In Fig. 3.1, the distance along the z-axis is divided into N equal parts, the uniform grid step size is Δz, and the spatial derivative first-order central difference scheme can be expressed as:

$$\left(\frac{\partial h}{\partial z}\right)_i = \frac{h^*_{i+1} - h^*_{i-1}}{2\Delta z} \tag{3.1}$$

where i represents the discrete grid nodes along the z-axis.

The spatial derivative second-order central difference format is expressed as:

Fig. 3.1 Uniform grid and Chebyshev grid

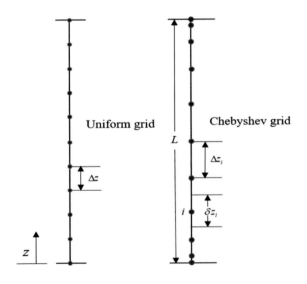

$$\left(\frac{\partial^2 h}{\partial z^2}\right)_i = \frac{h^*_{i+1} - 2h^*_i + h^*_{i-1}}{\Delta z^2} \tag{3.2}$$

Additionally, the first-order backward difference format of the time derivative is given by:

$$\left(\frac{\partial h}{\partial t}\right)_j = \frac{h^{*j} - h^{*j-1}}{\Delta t} \tag{3.3}$$

where Δt is the discrete time step, and j is the time node.

Substituting Eqs. (3.1)–(3.3) into the Richards' equation, one can obtain (Zhu et al. 2022a):

$$\left(\frac{h^{*j}_{i+1} - 2h^{*j}_i + h^{*j}_{i-1}}{\Delta z^2}\right) + \alpha\left(\frac{h^{*j}_{i+1} - h^{*j}_{i-1}}{2\Delta z}\right) = c\left(\frac{h^{*j}_i - h^{*j-1}_i}{\Delta t}\right) \quad 1 \le i \le N-1 \tag{3.4}$$

When the steady-state infiltration is incorporated, its finite difference format can be simplified to:

$$\left(\frac{h^*_{i+1} - 2h^*_i + h^*_{i-1}}{\Delta z^2}\right) + \alpha\left(\frac{h^*_{i+1} - h^*_{i-1}}{2\Delta z}\right) = 0 \tag{3.5}$$

The system of linear equations formed by Eqs. (3.4)–(3.5) can be further written in matrix form:

$$\mathbf{Ah^*} = \mathbf{b} \tag{3.6}$$

where $\mathbf{A}$ is a tridiagonal matrix of order $(N-1) \times (N-1)$; $\mathbf{h^*}$ and $\mathbf{b}$ are both column vectors of order $(N-1) \times 1$, and the first and last elements of vector $\mathbf{b}$ already contain boundary conditions. Equation (3.6) can be solved by relevant linear iterative methods, such as trigonometric decomposition method, Jacobi iterative method, Gauss–Seidel iterative method and relaxed iterative method.

3.2.2 Finite Volume Method

The finite volume method (FVM), also known as the control volume method, the basic idea is to divide the calculation area into a series of non-repetitive control volumes, and make a control volume around each grid point, and the differential equation is divided into a control volume. For each control volume integral, a system of linear equations to be solved is obtained. For the one-dimensional uniform grid coordinates in Fig. 3.1, the control volume here can be assumed to be $\Delta z \times 1 \times 1$, that is, the distance in the other directions except the z-axis is unit length. Furthermore, by integrating the Richards' equation over the uniform control volume, one can obtain:

$$\int_{i}^{i+\Delta z} \int_{t}^{t+\Delta t} \frac{\partial \theta}{\partial t} dt dz = \int_{t}^{t+\Delta t} \int_{i}^{i+\Delta z} \frac{\partial}{\partial z}\left[K_z(h)\left(\frac{\partial h}{\partial z}+1\right)\right] dz dt \qquad (3.7)$$

Furthermore, the discrete equation is written as:

$$\frac{K_{i+1/2}^{j}(h_{i+1}^{j} - h_{i}^{j})}{\Delta z} - \frac{K_{i-1/2}^{j}(h_{i}^{j} - h_{i-1}^{j})}{\Delta z} + K_{i+1/2}^{j} - K_{i-1/2}^{j}$$

$$= \Delta z C_{i}^{j-1/2} \frac{h_{i}^{j} - h_{i}^{j-1}}{\Delta t} \qquad (3.8)$$

where $K_{i+1/2}$ and $K_{i-1/2}$ are the harmonic mean values of the permeability coefficients of adjacent nodes:

$$K_{i+1/2} = \frac{2K_i K_{i+1}}{K_i + K_{i+1}} \qquad (3.9)$$

$$K_{i-1/2} = \frac{2K_i K_{i-1}}{K_i + K_{i-1}} \qquad (3.10)$$

Equation (3.8) can be further simplified into the following matrix form:

$$\mathbf{A}(\mathbf{h}^j)\mathbf{h}^j = \mathbf{b}(\mathbf{h}^j, \mathbf{h}^{j-1}) \qquad (3.11)$$

For the solution of Eq. (3.8), the Picard iteration method is usually used to solve it.

3.2.3 Finite Element Method

The finite element method (FEM) divides the solution domain into different elements, including triangular elements, rectangular elements, and quadrilateral elements. The equivalent integral equation of the problem can be obtained based on the variational principle and the orthogonalization principle of the weight function, and then the corresponding linear equation system can be obtained and solved iteratively. Usually, the Richards' equation solved by FEM is more than two-dimensional, then the two-dimensional linearized Richards' equation considering the x and z directions can be written as (Zhu et al. 2022b, c):

$$\frac{\partial^2 h^*}{\partial x^2} + \frac{\partial^2 h^*}{\partial z^2} + \alpha \frac{\partial h^*}{\partial z} = c \frac{\partial h^*}{\partial t} \qquad (3.12)$$

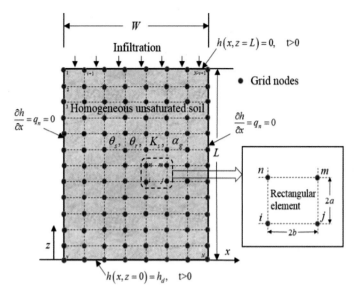

Fig. 3.2 2D infiltration model and rectangular element grid

As shown in Fig. 3.2, a rectangular element is used to divide the solution area. The number of discrete nodes along the x and z axes is N, then the discrete equation can be expressed in matrix form as follows:

$$\mathbf{Gh^*} = -\mathbf{P}\frac{\mathbf{h}^{*j} - \mathbf{h}^{*j-1}}{\Delta t} \tag{3.13}$$

where $\mathbf{G}$ represents the total conduction matrix on the left side of Eq. (3.12); $\mathbf{P}$ represents the storage matrix on the right side of Eq. (3.12); $\mathbf{h^*}$ is the array of the node pressure heads. Among them, the rectangular element (Fig. 3.2) composed of nodes i, j, m, and n only contributes to the Lth row of $\mathbf{G}$ ($L = i, j, m,$ and n). The Lth row elements of matrix $\mathbf{G}$ in a rectangular element can be given by:

$$G_{L,i}^e = \int_{-a}^{a}\int_{-b}^{b} \left(\frac{\partial N_i^e}{\partial x}\frac{\partial N_L^e}{\partial x} + \frac{\partial N_i^e}{\partial z}\frac{\partial N_L^e}{\partial z} - \alpha_g N_i^e N_L^e \right) \mathrm{d}x\,\mathrm{d}z \tag{3.14a}$$

$$G_{L,j}^e = \int_{-a}^{a}\int_{-b}^{b} \left(\frac{\partial N_j^e}{\partial x}\frac{\partial N_L^e}{\partial x} + \frac{\partial N_j^e}{\partial z}\frac{\partial N_L^e}{\partial z} - \alpha_g N_j^e N_L^e \right) \mathrm{d}x\,\mathrm{d}z \tag{3.14b}$$

$$G_{L,m}^e = \int_{-a}^{a}\int_{-b}^{b} \left(\frac{\partial N_m^e}{\partial x}\frac{\partial N_L^e}{\partial x} + \frac{\partial N_m^e}{\partial z}\frac{\partial N_L^e}{\partial z} - \alpha_g N_m^e N_L^e \right) \mathrm{d}x\,\mathrm{d}z \tag{3.14c}$$

$$G_{L,n}^e = \int\limits_{-a}^{a} \int\limits_{-b}^{b} \left(\frac{\partial N_n^e}{\partial x} \frac{\partial N_L^e}{\partial x} + \frac{\partial N_n^e}{\partial z} \frac{\partial N_L^e}{\partial z} - \alpha_g N_n^e N_L^e \right) dx\,dz \qquad (3.14d)$$

where N_L^e represents the element basis function; and a and b are half the length and width of the rectangular element, respectively. N_L^e can be formatted as:

$$N_i^e(x, z) = \frac{1}{4}\left(1 - \frac{x}{b}\right)\left(1 - \frac{z}{a}\right) \qquad (3.15a)$$

$$N_j^e(x, z) = \frac{1}{4}\left(1 + \frac{x}{b}\right)\left(1 - \frac{z}{a}\right) \qquad (3.15b)$$

$$N_m^e(x, z) = \frac{1}{4}\left(1 + \frac{x}{b}\right)\left(1 + \frac{z}{a}\right) \qquad (3.15c)$$

$$N_n^e(x, z) = \frac{1}{4}\left(1 - \frac{x}{b}\right)\left(1 + \frac{z}{a}\right) \qquad (3.15d)$$

Similarly, the Lth row elements of $\mathbf{P}$ can also be formed as:

$$P_{L,i}^e = c \int\limits_{-a}^{a} \int\limits_{-b}^{b} N_i^e N_L^e dx\,dz \qquad (3.16a)$$

$$P_{L,j}^e = c \int\limits_{-a}^{a} \int\limits_{-b}^{b} N_j^e N_L^e dx\,dz \qquad (3.16b)$$

$$P_{L,m}^e = c \int\limits_{-a}^{a} \int\limits_{-b}^{b} N_m^e N_L^e dx\,dz \qquad (3.16c)$$

$$P_{L,n}^e = c \int\limits_{-a}^{a} \int\limits_{-b}^{b} N_n^e N_L^e dx\,dz \qquad (3.16d)$$

Furthermore, the contributions of all rectangular elements are summed and assembled to form matrices $\mathbf{G}$ and $\mathbf{P}$. Generally, Eqs. (3.14)–(3.16) are solved using Gaussian integration method (Wang and Anderson 1982). The Gaussian integration method makes the definite integral equal to a weighted sum over a finite number of points, and for a quadratic polynomial, the general expression for the double integral is as follows:

$$\int_{-1}^{1}\int_{-1}^{1} f(\varepsilon, \eta)\mathrm{d}\varepsilon\mathrm{d}\eta = f(\varepsilon_1, \eta_1) + f(\varepsilon_2, \eta_1) + f(\varepsilon_1, \eta_2) + f(\varepsilon_2, \eta_2) \quad (3.17)$$

Among them, the four Gauss points are determined by $\varepsilon_1 = -1/\sqrt{3}$, $\varepsilon_2 = 1/\sqrt{3}$, $\eta_1 = -1/\sqrt{3}$, $\eta_2 = 1/\sqrt{3}$, and the ownership coefficients are all equal to 1. The integral of Eqs. (3.14)–(3.16) can be transformed into the form of Eq. (3.17) by the following variable transformation:

$$\varepsilon = \frac{x}{b} \text{ and } \eta = \frac{z}{a} \quad (3.18)$$

Then, $\mathrm{d}z = a\mathrm{d}\varepsilon$, $\mathrm{d}x = b\mathrm{d}\eta$, the integral limit is from -1 to 1, and the interpolation function (Eq. 3.15) in the coordinates (ε, η) is transformed into:

$$\overline{N}_i^e(\varepsilon, \eta) = \frac{1}{4}(1 - \varepsilon)(1 - \eta) \quad (3.19a)$$

$$\overline{N}_j^e(\varepsilon, \eta) = \frac{1}{4}(1 + \varepsilon)(1 - \eta) \quad (3.19b)$$

$$\overline{N}_m^e(\varepsilon, \eta) = \frac{1}{4}(1 + \varepsilon)(1 + \eta) \quad (3.19c)$$

$$\overline{N}_n^e(\varepsilon, \eta) = \frac{1}{4}(1 - \varepsilon)(1 + \eta) \quad (3.19d)$$

In summary, the integral transformation of Eqs. (3.14)–(3.16) is described as:

$$\int_{-a}^{a}\int_{-b}^{b} f(x, y)\mathrm{d}x\mathrm{d}y = ab\int_{-1}^{1}\int_{-1}^{1} f(\varepsilon, \eta)\mathrm{d}\varepsilon\mathrm{d}\eta \quad (3.20)$$

where $f(\varepsilon, \eta)$ is the transformation of $f(x, z)$ at the coordinate (ε, η).

3.2.4 Numerical Approximation to the Fractional-Time Richards' Equation

Fractional Richards' equations can be further divided into space fractional, time fractional, and space–time fractional partial differential equations. The time fractional Richards' equation is expressed as follows:

$$C(h)\frac{\partial^\gamma h}{\partial t^\gamma} = \frac{\partial}{\partial z}\left[K(h)\left(\frac{\partial h}{\partial z} + 1\right)\right] \quad (3.21)$$

where γ is the order of the time derivative. When fractional order $\gamma = 1$, Eq. (3.21) degenerates into the classical Richard equation.

The Caputo fractional derivative with respect to the function $f(x)$ can be defined as (Pachepsky et al. 2003):

$$
{}_{0}^{C}D_{x}^{\gamma}f(x) = \frac{1}{\Gamma(n-\gamma)} \int_{0}^{x} \frac{f^{(n)}(\tau)}{(x-\tau)^{\gamma-n+1}} \, d\tau \tag{3.22}
$$

where ${}_{0}^{C}D_{x}^{\gamma}$ represents the γ-order Caputo fractional derivative; Γ is the gamma function, and n is a positive integer (representing an integer derivative). In addition, the Riemann–Liouville (R-L) fractional derivative can be defined as (Su 2014):

$$
{}_{0}^{RL}D_{x}^{\gamma}f(x) = \frac{1}{\Gamma(n-\gamma)} \frac{d^{n}}{dx^{n}} \int_{0}^{x} \frac{f(\tau)}{(x-\tau)^{\gamma-n+1}} \, d\tau \tag{3.23}
$$

where ${}_{0}^{RL}D_{x}^{\gamma}$ represents the γ-order R-L fractional derivative.

It can be seen from Eqs. (3.22)–(3.23) that in the definition of R-L fractional derivative, the fractional integral is first obtained and then the integer derivative is obtained, while the definition of Caputo derivative is to first obtain the integer derivative, and then the fractional integral is calculated. There is a great connection between the two, and there are also essential differences, which are mainly reflected in the actual physical models. In the process of actually solving the initial value problem of differential equations, Caputo derivatives are more widely used and have more physical background than R-L derivatives.

For the solution of Eq. (3.21), the finite difference method is used for numerical discretization, and meshing is conducted. Let η and Δz be the time and space steps, respectively. The simulation time is divided into M equal parts, and the z-axis is divided into N equal parts:

$$
t_{m} = m\eta, \quad m = 0, 1, \ldots M \tag{3.24}
$$

$$
z_{i} = i\Delta z, \quad i = 0, 1, \ldots N \tag{3.25}
$$

For the left-hand fractional derivative term of Eq. (3.21), the Caputo fractional derivative is defined as:

$$
\frac{\partial^{\gamma}h(z,t)}{\partial t^{\gamma}} = \frac{1}{\Gamma(1-\gamma)} \int_{0}^{t} \frac{\partial h(z,\tau)}{\partial \tau} \frac{d\tau}{(t-\tau)^{\gamma}} \tag{3.26}
$$

The integral of the pressure head derivative in Eq. (3.26) can be directly approximated by the numerical differential equation, which is deduced as follows:

$$\frac{\partial^\gamma h(z, t_m)}{\partial t^\gamma} = \frac{1}{\Gamma(1-\gamma)} \int_0^{t_m} \frac{h'(\tau)d\tau}{(t_m - \tau)^\gamma} = \frac{1}{\Gamma(1-\gamma)} \sum_{j=0}^{m-1} \int_{t_j}^{t_{j+1}} \frac{h'(t_m - \tau)d\tau}{\tau^\gamma}$$

$$\approx \frac{1}{\Gamma(1-\gamma)} \sum_{j=0}^{m-1} \frac{h(t_m - t_j) - h(t_m - t_{j+1})}{\eta} \int_{t_j}^{t_{j+1}} \tau^{-\gamma} d\tau$$

$$= \frac{\eta^{-\gamma}}{\Gamma(2-\gamma)} \sum_{j=0}^{m-1} (h_{m-j} - h_{m-j-1})[(j+1)^{1-\gamma} - j^{1-\gamma}] \qquad (3.27)$$

Equation (3.27) is further simplified as:

$$\frac{\partial^\gamma h(z, t_m)}{\partial t^\gamma} = \frac{\eta^{-\gamma}}{\Gamma(2-\gamma)} \sum_{j=0}^{m-1} b_j^{(\gamma)} (h_{m-j} - h_{m-j-1}) \qquad (3.28)$$

where $b_j^{(\gamma)} = (j+1)^{1-\gamma} - j^{1-\gamma}$.

For finite difference discretization on the right side of Eq. (3.21), one obtains:

$$\frac{\partial}{\partial z}\left[K(h)\left(\frac{\partial h}{\partial z} + 1\right)\right] = \frac{K_{i+1/2}^{m+1}(h_{i+1}^{m+1} - h_i^{m+1})}{\Delta z^2} - \frac{K_{i-1/2}^{m+1}(h_i^{m+1} - h_{i-1}^{m+1})}{\Delta z^2}$$

$$+ \frac{K_{i+1/2}^{m+1} - K_{i-1/2}^{m+1}}{\Delta z} \qquad (3.29)$$

$K_{i+1/2}$ and $K_{i-1/2}$ can be expressed as:

$$K_{i+1/2} = \frac{2 K_i K_{i+1}}{K_i + K_{i+1}} \qquad (3.30)$$

$$K_{i-1/2} = \frac{2 K_i K_{i-1}}{K_i + K_{i-1}} \qquad (3.31)$$

Combining Eqs. (3.28) and (3.29), the discrete format of the fractional-time Richards' equation can be obtained as follows:

$$\frac{\eta^{-\gamma}}{\Gamma(2-\gamma)} C(h)_i^{m-1/2} \sum_{j=0}^{m-1} b_j^{(\gamma)} (h_{m-j,i} - h_{m-j-1,i})$$

$$= \frac{K_{i+1/2}^m (h_{i+1}^m - h_i^m)}{\Delta z^2} - \frac{K_{i-1/2}^m (h_i^m - h_{i-1}^m)}{\Delta z^2} + \frac{K_{i+1/2}^m - K_{i-1/2}^m}{\Delta z} \qquad (3.32)$$

where $C(h)_i^{m-1/2}$ represents the average value of the specific moisture capacity $(C(h)_i^m)$ at the previous time step and the specific moisture capacity $(C(h)_i^{m-1,k})$ of the current iteration step $(k \geq 1)$ of the current time step.

Equation (3.32) can be simplified into matrix form $\mathbf{Ax} = \mathbf{b}$, and then iteratively solved by Picard method. In order to evaluate the fitting effect of the proposed time fractional model, two indicators are selected, namely the root mean square error (RMSE) and the relative error (RE):

$$
\text{RSE} = \left[\frac{1}{N-1} \sum_{i=1}^{N-1} \left(\frac{h_i - h_i^*}{h_i^*} \right)^2 \right]^{0.5} \tag{3.33}
$$

$$
\text{RE} = 100 \times \left[\frac{1}{N-1} \sum_{i=1}^{N-1} \left(\left| \frac{h_i - h_i^*}{h_i^*} \right| \right) \right] \tag{3.34}
$$

where h_i is the numerical solution of Richards' equation, and h_i^* is the exact solution of Richards' equation. The smaller the values of the two errors, the higher the computational accuracy of the proposed approach.

3.3 An Improved Method for Non-uniform Spatial Grid

In the numerical solution of Richards' equation, FDM can be used for numerical discretization and iterative solution. However, to obtain a more reliable numerical solution, the space step size of conventional uniform grid (Fig. 3.1) is often small, particularly under some unfavorable numerical conditions, such as infiltration into dry and layered soils with greatly different permeability coefficients, this makes the calculation time-consuming and even the accuracy cannot be improved very well. The realization of unstructured space grids and dynamic grid methods in some studies is often complicated and inappropriate (Chávez-Negrete et al. 2018; Dolejší et al. 2019). Therefore, this chapter proposes an improved FDM numerical discretization process using a non-uniform Chebyshev space grid, which is compared with the traditional uniform space grid. This method can provide a certain reference for the numerical simulation of unsaturated infiltration.

(1) Uniform grid method

The Richards' equation is directly numerically discretized by the finite difference method of uniform grid, and one can gain (Zhu et al. 2020):

$$
\frac{\left[K_{i+1/2}^j (h_{i+1}^j - h_i^j) - K_{i-1/2}^j (h_i^j - h_{i-1}^j) \right]}{\Delta z^2} + \frac{K_{i+1/2}^j - K_{i-1/2}^j}{\Delta z}
$$
$$
= \frac{C_i^{j-1/2} \left(h_i^j - h_i^{j-1} \right)}{\Delta t} \tag{3.35}
$$

Comparing Eq. (3.29), it can be found that the mathematical meaning of Eqs. (3.35) and (3.29) is consistent. When its steady-state infiltration is considered, the FDM format can be simplified as (Zhu et al. 2020):

$$\frac{\left[K_{i+1/2}(h_{i+1} - h_i) - K_{i-1/2}(h_i - h_{i-1})\right]}{\Delta z^2} + \frac{K_{i+1/2} - K_{i-1/2}}{\Delta z} = 0 \qquad (3.36)$$

(2) Improved Chebyshev space grid method

Because the pressure head in the unsaturated infiltration problem often has a large change at the boundary and the soil layer interface, a finer mesh is usually required for densification. However, the uniform grid generates too many computing grid nodes in the process of densification, and the computation is time-consuming and even not accurate enough. Then, a Chebyshev grid coordinate is proposed as (Wu et al. 2020b; Zhu et al. 2020):

$$z_i = \cos(i\pi/N) \times \frac{L}{2} + \frac{L}{2}, \quad i = N, N - 1, \ldots, 0 \qquad (3.37)$$

where L is the thickness of the soil layer.

In Fig. 3.1, the Chebyshev grid nodes are only highly refined at the interface, greatly reducing the number of grid nodes. Furthermore, the FDM discrete format combined with the Chebyshev grid can be expressed as:

$$\frac{1}{\delta z_i}\left[\frac{K_{i+1/2}^j(h_{i+1}^j - h_i^j)}{\Delta z_{i+1}} - \frac{K_{i-1/2}^j(h_i^j - h_{i-1}^j)}{\Delta z_i}\right] + \frac{K_{i+1/2}^j - K_{i-1/2}^j}{\delta z_i}$$

$$= C_i^{j-1/2}\frac{h_i^j - h_i^{j-1}}{\Delta t} \qquad (3.38)$$

where Δz_i represents the unequal spacing between the Chebyshev grid nodes; δz_i is the unequal spacing of the computing nodes (Fig. 3.1).

Equation (3.38) can also be simplified into a matrix form such as Eq. (3.13). Due to the nonlinear relationship between permeability coefficient and water content, the coefficient matrix **A** needs to be repeatedly evaluated by nonlinear iteration method after numerical discretization. Among them, Picard method is a more classical and practical nonlinear iterative method. Based on the Chebyshev grid discretization format of the Richards' equation, a program for unsaturated infiltration was developed using the MATLAB (R2014a) language.

3.3.1 Validation Example

This test describes one-dimensional transient unsaturated infiltration in homogeneous unsaturated soil (Ku et al. 2018; Zhu et al. 2022a), the soil thickness $L = 10$ m, exponential model parameters are: $\alpha = 1 \times 10^{-4}$, $\theta_s = 0.50$, $\theta_r = 0.11$, and the saturated permeability coefficient $k_s = 2.5 \times 10^{-8}$ m/s. Furthermore, the boundary conditions can be given by:

$$h(z = 0) = h_d \tag{3.39}$$

$$h(z = 10) = 0 \tag{3.40}$$

To verify the computational accuracy of the proposed method, the following error indicators are used in the comparative analysis, namely the root mean square error (RSE), the relative error (RE), and maximum relative error (MRE):

$$RSE = \left[\frac{1}{N-1} \sum_{i=1}^{N-1} \left(\frac{h_i - h_i^*}{h_i^*} \right)^2 \right]^{0.5} \tag{3.41}$$

$$RE = 100 \times \left[\frac{1}{N-1} \sum_{i=1}^{N-1} \left(\left| \frac{h_i - h_i^*}{h_i^*} \right| \right) \right] \tag{3.42}$$

$$MRE = 100 \times \left[\max \left| \frac{h_i - h_i^*}{h_i^*} \right|_{i=1}^{N-1} \right] \tag{3.43}$$

where h_i is the numerical solution of Richards' equation and h_i^* is the exact solution of Richards' equation. The smaller the values of the three errors, the higher the computational accuracy of the proposed method. The total simulation time was set to 5 h, the number of discrete nodes was set to 100, 150, and 200, and the time steps were set to 0.01, 0.02, and 0.04 h, respectively.

In Fig. 3.3, the numerical solutions are derived using two grid methods under the conditions of $\Delta t = 0.01$ h and $N = 200$ and compared with the analytical solutions. The numerical solution obtained by the uniform grid method has a large deviation from the exact solution, particularly after $t > 2$ h (Fig. 3.3a), while the numerical solution obtained by the Chebyshev grid method is in good agreement with the analytical solution (Fig. 3.3b).

Figure 3.4a represents the maximum relative error (MRE) obtained by different grid methods at different time steps when the number of nodes N is 100. The MRE obtained by the Chebyshev grid method ranges from 0.6 to 3.5%, and it first decreases and then increases with the increase of time. When $t < 4$ h, the MRE decreases with decreasing time step. The MRE obtained by the uniform grid method ranges from 1.3 to 49%, particularly when $t > 1$ h, the MRE increases over time and is much larger than that obtained by the Chebyshev grid.

Figure 3.4b shows the MRE obtained by different grid methods under different numbers of discrete grid nodes when the time step $\Delta t = 0.01$ h. It can be found that the MRE obtained by the uniform grid as time increases. In addition, the MRE of the two methods has a decreasing trend with the increase of the number of grid nodes N.

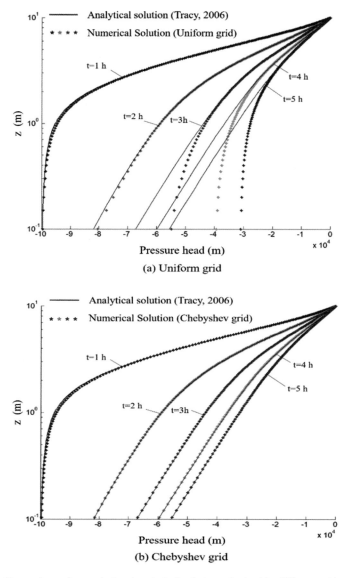

Fig. 3.3 Comparison of numerical and analytical solutions obtained by different grid methods

In Table 3.2, the RSE of both methods and the RE of the uniform grid decrease with the increase of N, while the RE of the Chebyshev grid increases with increasing N. However, it can be seen from the numerical value that the RSE of the Chebyshev grid is nearly 100 times different from the uniform grid method, and the RE is more than 10 times different from the uniform grid method. This test indicates that the proposed Chebyshev grid method is not limited by the number of grid nodes

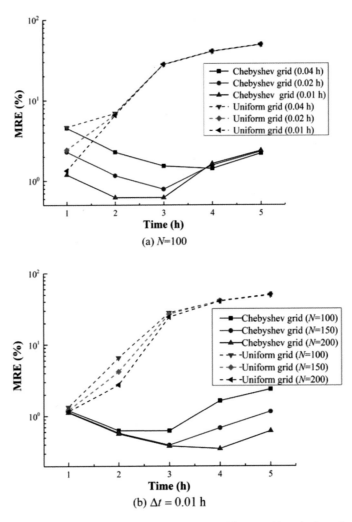

Fig. 3.4 Comparison of the maximum relative error of different grid methods under different numerical conditions

to improve the accuracy, that is, the proposed method achieves higher numerical accuracy with fewer discrete nodes, and has a smaller computational cost.

3.3.2 Unsaturated Infiltration in Layered Soils

The mathematical model is shown in Fig. 3.5. The model parameters of the two-layer soils are set to $\theta_s = 0.35$, $\theta_r = 0.14$, $\alpha = 8 \times 10^{-3}$. The thickness of soil layer 1 and

Table 3.2 Numerical accuracy at $t = 5$ h

Condition	RSE		RE (%)	
N	Uniform grid	Chebyshev grid	Uniform grid	Chebyshev grid
100	0.11	3.7e−3	5.01	0.050
150	0.10	2.6e−3	3.53	0.122
200	0.09	2.1e−3	3.05	0.148

soil layer 2 are both set to 5 m, the number of discrete nodes in each layer is 40, and the boundary conditions are consistent with Sect. 3.3.1, where $h_d = -10^3$ m.

In layered soils, due to the influence of different soil types, different combinations have a great influence on unsaturated infiltration. Table 3.3 lists the saturated permeability coefficients in different saturated soils. In order to further verify the applicability of the proposed grid method, it is assumed that the saturated permeability coefficient of soil layer 1 is 10^{-1} m/s, and the saturated permeability coefficient of the second layer gradually changes from coarse sandy soil to clay in Table 3.3, the saturated permeability coefficient of the second layer ranges from 10^{-2} to 10^{-9} m/s.

Fig. 3.5 Chebyshev grid of two layers in unsaturated soil

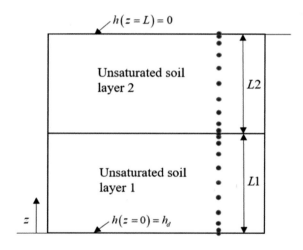

Table 3.3 Typical permeability coefficient values for saturated soils

Soil type	Permeability coefficient (m/s)
Clean gravel	$1-10^{-2}$
Coarse sand	$10^{-2}-10^{-4}$
Fine sand	$10^{-4}-10^{-5}$
Silty sand	$10^{-5}-10^{-7}$
Clay	$< 10^{-8}$

It can be seen from Fig. 3.6a that the numerical solutions obtained by the uniform grid and the Chebyshev grid are approximate, and both can well simulate the unsaturated infiltration process in case 2. Figure 3.6b depicts the numerical results of case 4. The ratio of the saturated permeability coefficients of the upper and lower soil layers is in the order of 10^4. The uniform grid cannot accurately represent the pressure head at the interface, but the numerical solution obtained by the Chebyshev grid can accurately describe the pressure head variation at the interface. In Fig. 3.6c, the Chebyshev grid can also accurately describe the variations in the pressure head at the interface under case 6 (Table 3.4).

Similar to the two-layer soil, the proposed improved grid method is applied to the three-layer unsaturated soils. The mathematical model is described in Fig. 3.7, where $L_1 = L_3 = 4$ m and $L_2 = 2$ m. The number of discrete nodes in each layer of soil is 40, the parameter θ_s is set to 0.46, the three-layer unsaturated soil permeability coefficient is listed in Table 3.5, and the boundary conditions can be expressed as: $h(z = 0) = 0$, $h(z = 10 \text{ m}) = -1000$ m. Other numerical conditions are consistent with the two-layered soils.

In Fig. 3.8, the proposed Chebyshev grid method can better characterize the pressure head change between the two interfaces in case 11 compared to the uniform grid method. This test further demonstrates that the proposed Chebyshev grid method can obtain more reliable numerical solutions with fewer grid nodes under some unfavorable infiltration conditions, particularly when the saturated permeability coefficient changes greatly in layered soils.

3.4   Application of Chebyshev Spectral Method to Richards' Equation

3.4.1   Chebyshev Spectral Method

The Chebyshev spectral method (CSM) was developed by Gottlieb et al. (1978). It is widely used to solve numerical problems expressed by partial differential equations. The advantage of the CSM is the use of non-uniform mesh discretization and the Chebyshev differentiation matrix to improve the accuracy of the numerical simulation. In Fig. 1.2, the $N + 1$ Chebyshev point coordinates in the interval $[-1, 1]$ can be expressed as:

$$z_j = \cos(j\pi/N), \quad j = 0, 1, \ldots N \qquad (3.44)$$

The first-order derivative of interpolation function $p'(\mathbf{z})$ can be written as:

$$p'(\mathbf{z}) = \mathbf{D}_N \mathbf{h} \qquad (3.45)$$

Fig. 3.6 Comparison of numerical solutions under different cases

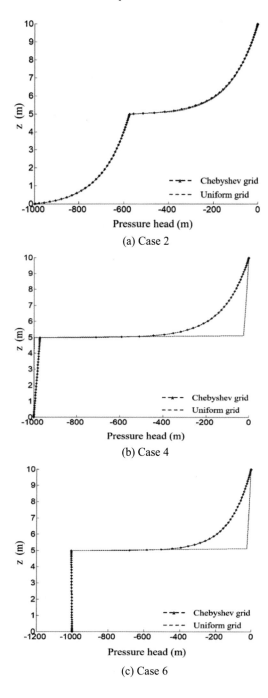

(a) Case 2

(b) Case 4

(c) Case 6

Table 3.4 Permeability coefficient for cases 1–8

Case number	1	2	3	4	5	6	7	8
K_{s1} (m/s)	10^{-1}	10^{-1}	10^{-1}	10^{-1}	10^{-1}	10^{-1}	10^{-1}	10^{-1}
K_{s2} (m/s)	10^{-2}	10^{-3}	10^{-4}	10^{-5}	10^{-6}	10^{-7}	10^{-8}	10^{-9}

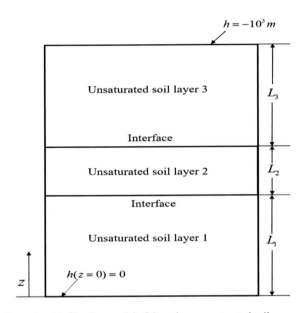

Fig. 3.7 One-dimensional infiltration model of three-layer unsaturated soil

Table 3.5 Permeability coefficient for cases 9–13

Case number	9	10	11	12	13
K_{s1} (m/s)	1	1	1	1	1
K_{s2} (m/s)	10^{-3}	10^{-4}	10^{-5}	10^{-6}	10^{-7}
K_{s3} (m/s)	1	1	1	1	1

where $\mathbf{z}$ is represented as the $N + 1$-dimensional vector $(z_0, z_1, \ldots, z_N)^{\mathrm{T}}$, $\mathbf{h}$ represents the $N + 1$ dimensional vector $(h_0, h_1, \ldots, h_N)^{\mathrm{T}}$ composed of the pressure head, and $\mathbf{D}_N$ is an $(N + 1) \times (N + 1)$ Chebyshev differentiation matrix.

The expression of each element in Chebyshev differentiation matrix $\mathbf{D}_N$ for any number of nodes $(N + 1)$ is given by:

$$(\mathbf{D}_N)_{00} = \frac{2N^2 + 1}{6}, \quad (\mathbf{D}_N)_{NN} = -\frac{2N^2 + 1}{6}, \tag{3.46a}$$

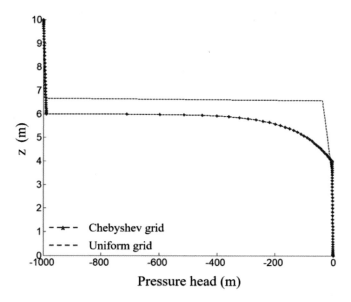

Fig. 3.8 Comparison of numerical solutions for case 11

$$(\mathbf{D}_N)_{jj} = \frac{-z_j}{2\left(1 - z_j^2\right)}, \quad j = 1, \ldots, N - 1 \tag{3.46b}$$

$$(\mathbf{D}_N)_{ij} = \frac{c_i}{c_j} \frac{(-1)^{i+j}}{z_i - z_j}, \quad (i \neq j)i, j = 0, \ldots, N, \tag{3.46c}$$

where $(\mathbf{D}_N)_{ij}$ represents the element in row $i + 1$ and column $j + 1$ of Chebyshev differentiation matrix $\mathbf{D}_N$ for the first derivative and,

$$c_i = \begin{cases} 2, \ i = 0, N \\ 1, \ i = 1, \ldots, N - 1 \end{cases} \tag{3.47}$$

To obtain the Chebyshev differentiation matrix for the second derivative, the matrix can be directly squared to $\mathbf{D}_N$. Similarly, Chebyshev differentiation matrices for higher-order derivatives can be obtained as follows:

$$\frac{\partial}{\partial z} \rightarrow \mathbf{D}_N \tag{3.48a}$$

$$\frac{\partial^2}{\partial z^2} \rightarrow \mathbf{D}_N^2 \tag{3.48b}$$

$$\frac{\partial^3}{\partial z^3} \rightarrow \mathbf{D}_N^3 \tag{3.48c}$$

Because $\mathbf{D}_N$ is obtained in the interval $[-1, 1]$, it is necessary to scale $\mathbf{D}_N$ to other intervals, that is,

$$\frac{\partial}{\partial z} \rightarrow \frac{2}{L}\mathbf{D}_N \tag{3.49a}$$

$$\frac{\partial^n}{\partial z^n} \rightarrow \left(\frac{2}{L}\mathbf{D}_N\right)^n \tag{3.49b}$$

Through the above process, the partial differential equation including the RE can be easily and efficiently solved. Thus, the governing equation (Eq. 2.1) can be rewritten as:

$$\frac{1}{c}\mathbf{D}_N^2\mathbf{h}^* + \frac{\alpha_g}{c}\mathbf{D}_N\mathbf{h}^* = \frac{\partial\mathbf{h}^*}{\partial t} \tag{3.50}$$

In Eq. (3.50), Eq. (2.1) has been transformed into an ordinary differential equation (ODE) by introducing Chebyshev differentiation matrices. In this book, the variable-step four- and fifth-order Runge–Kutta method is used to solve Eq. (3.50), which is the ODE solver in MATLAB. Meanwhile, the 2D linear RE can also be solved using the same method.

To evaluate the performance of the proposed method, an L_2 norm is used to indicate the error of solution h^*:

$$L_2(h^*) = \left\|h^{*,a}(z, t) - h^*(z, t)\right\|_2 \tag{3.51}$$

where $h^{*,a}(z, t)$ and $h^*(z, t)$ are one-dimensional (1D) analytical and numerical solutions, respectively. Simultaneously, the numerical and analytical solutions with spatial and temporal grids can produce an absolute error defined as follows:

$$\mathrm{err}_A(h^*) = h^{*,a}(z, t) - h^*(z, t). \tag{3.52}$$

3.4.2 Validation Example

Test 1 represents transient infiltration in homogeneous unsaturated soils. In the mathematical model, the soil thickness (L) is assumed to be 10 m. The model parameters are $\theta_s = 0.50$, $\theta_r = 0.11$, $\alpha = 1 \times 10^{-4}$, and $k_s = 9 \times 10^{-5}$ m/h (Zhu et al. 2019). The governing equation is described as Eq. (2.1).

When water infiltrates into a soil mass, ponding on the ground maintains the pressure head at zero (Green and Ampt 1911). The upper and lower boundary conditions can be expressed as:

$$h(z = 0, t) = h_d \tag{3.53}$$

$$h(z = L, t) = 0 \tag{3.54}$$

The effect of Δz and N on the results of Test 1 was analyzed. The time step was set to 0.01 h. Different numbers of nodes or grid sizes, that is, $N/\Delta z = 20/0.5$ (m), $N/\Delta z = 40/0.25$ (m), and $N/\Delta z = 80/0.125$ (m) were selected to investigate the influence of the different methods on $L_2(h^*)$. The total simulation time was 5 h. Simultaneously, three different initial conditions h_0 were chosen to evaluate their influence on the numerical accuracy and stability of the FDM and CSM.

In Fig. 3.9, $L_2(h^*)$ of the CSM were consistently smaller than that of the conventional FDM. The numerical accuracy increased as the mesh was refined, and the accuracy of the traditional FDM was significantly affected by the initial conditions. However, the proposed method (CSM) was less affected, and the numerical accuracy was stable in the order of 10^{-6}–10^{-7} in this example. The result indicates that the CSM was more robustness than the FDM. Figure 3.10 demonstrates the comparison of the computed pressure head profiles with different grid sizes at $t = 2$ h. Compared with the analytical solution, it can be easily seen that the numerical accuracy of the CSM was less influenced by the grid sizes than that of the FDM (Fig. 3.10). That is, the CSM achieved better numerical results with fewer mesh nodes.

The pressure head profiles were obtained from the computed results over time under $h_0 = -100,000$ m (Fig. 3.11). The number of nodes or grid sizes was 40/0.25 (m). Figure 3.11 illustrates that the numerical solutions of the CSM agree well with the analytical solutions over time, whereas the numerical solutions of the FDM have larger errors than the analytical solutions. In Fig. 3.12, the minimum absolute errors of the proposed CSM and the FDM are approximately 10^{-11} and 10^{-6}, respectively. Consequently, the extended CSM is characterized by higher accuracy than traditional FDM with fewer nodes.

3.5 Conclusions

This chapter first summarizes the commonly used spatial numerical discrete methods of Richards' equation, including FDM, FVM, and FEM. Furthermore, this chapter proposes an improved FDM numerical discretization process using a non-uniform Chebyshev space grid and compares it with the numerical results and analytical solutions obtained from a conventional uniform space grid. Additionally, based on the non-uniform Chebyshev grid method, the Chebyshev spectral method is proposed to solve the Richards' equation, and the following conclusions can be obtained:

(1) The proposed Chebyshev grid method, while keeping the number of discrete nodes unchanged, uses a cosine function to ingeniously densify the interfaces on both sides and then numerically solve it to obtain a more reliable numerical solution. The numerical results indicate that the numerical solutions obtained

Fig. 3.9 Comparison of the
calculation accuracy for
solving 1D transient
infiltration with different grid
sizes and initial conditions
using different methods: **a** h_0
$= -10$ m; **b** $h_0 = -$
1000 m; **c** $h_0 = -100{,}000$ m

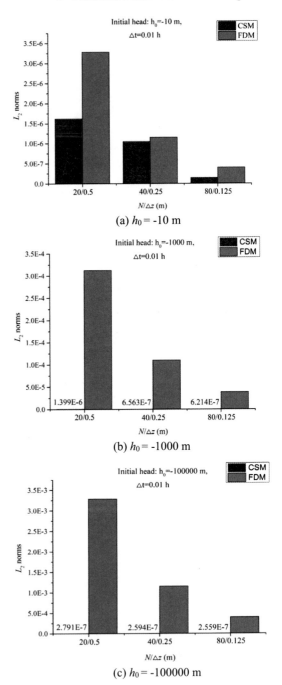

(a) $h_0 = $ -10 m

(b) $h_0 = $ -1000 m

(c) $h_0 = $ -100000 m

Fig. 3.10 Comparison of the computed profiles of the pressure head with different grid sizes at $t = 2$ h: **a** $\Delta z = 0.5$ m; **b** $\Delta z = 0.25$ m; **c** $\Delta z = 0.125$ m

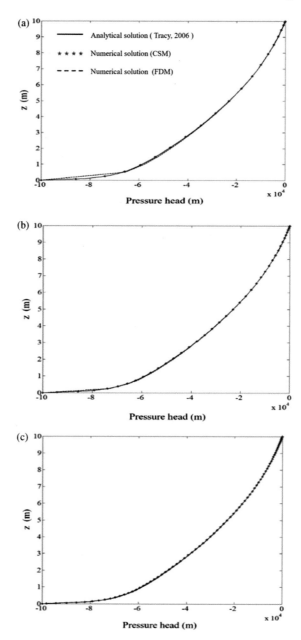

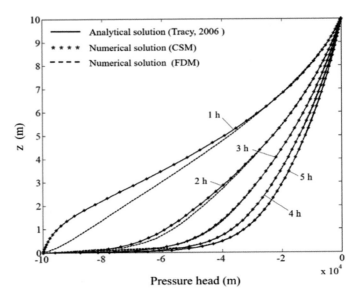

Fig. 3.11 Comparison of the calculation results (pressure head) from different methods when h_0 $= -100,000$ m

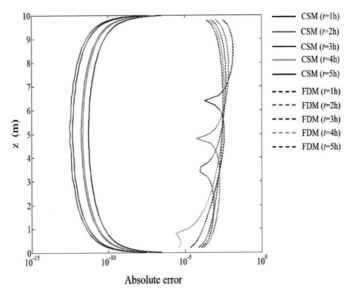

Fig. 3.12 Absolute error in the results computed using the CSM and FDM relative to the analytical solution for the 1D transient infiltration at different simulated times

by this method are in good agreement with the analytical solutions under some unfavorable numerical conditions, such as infiltration into a dry soil. At the same time, the results illustrate that the proposed Chebyshev grid method can obtain higher numerical accuracy with a smaller number of nodes than the conventional uniform grid, and the computational cost is small.

(2) The Chebyshev spectral method (CSM) based on the discretization of Chebyshev grid is extended to simulate unsaturated infiltration in porous media. Chebyshev differential matrix can construct n-order Chebyshev differential matrix conveniently and quickly. This is the first time that the CSM has been used successfully to solve 1D and 2D transient infiltration problems in unsaturated soils. Compared with the traditional FDM, the CSM was more efficient, and it achieved higher accuracy with a lower-resolution grid. The CSM was not sensitive to the initial conditions.

References

An HY, Ichikawa Y, Tachikawa Y et al (2010) Three-dimensional finite difference saturated-unsaturated flow modeling with nonorthogonal grids using a coordinate transformation method. Water Resour Res 46:W11521

Bergamaschi L, Putti M (1999) Mixed finite elements and Newton-type linearizations for the solution of Richards' equation. Int J Numer Meth Eng 45(8):1025–1046

Broadbridge P, Daly E, Goard J (2017) Exact solutions of the Richards equation with nonlinear plant-root extraction. Water Resour Res 53(11):9679–9691

Chávez-Negrete C, Domínguez-Mota FJ, Santana-Quinteros D (2018) Numerical solution of Richards' equation of water flow by generalized finite differences. Comput Geotech 101:168–175

Crevoisier D, Chanzy A, Voltz M (2009) Evaluation of the Ross fast solution of Richards' equation in unfavourable conditions for standard finite element methods. Adv Water Resour 32:936–947

Cross D, Onof C, Winter H (2020) Ensemble estimation of future rainfall extremes with temperature dependent censored simulation. Adv Water Resour 136:103479

Dolejší V, Kuraz M, Solin P (2019) Adaptive higher-order space-time discontinu-ous Galerkin method for the computer simulation of variably-saturated porous media flows. Appl Math Model 72: 276-305

Eini N, Afshar MH, Faraji Gargari S et al (2020) A fully Lagrangian mixed discrete least squares meshfree method for simulating the free surface flow problems. Eng Comput. https://doi.org/10.1007/s00366-020-01157-x

Eymard R, Hilhorst D, Vohralik M (2006) A combined finite volume-nonconforming/mixed-hybrid finite element scheme for degenerate parabolic problems. Numer Math 105:73–131

Fahs M, Younes A, Lehmann F (2009) An easy and efficient combination of the mixed finite element method and the method of lines for the resolution of Richards' equation. Environ Model Softw 24:1122–1126

Gottlieb D, Orszag SA, Sod GA (1978) Numerical analysis of spectral methods: theory and application (CBMS-NSF regional conference series in applied mathematics). J Appl Mech 45(4):969

Green WH, Ampt GA (1911) Studies on soil physics I. The flow of air and water through soils. Journal of Agricultural Research 4(1), 1–24

Herrera PA, Massabó M, Beckie RD (2009) A meshless method to simulate solute transport in heterogeneous porous media. Adv Water Resour 32(3):413–429

Jiang SH, Liu X, Huang J (2022) Non-intrusive reliability analysis of unsaturated embankment slopes accounting for spatial variabilities of soil hydraulic and shear strength parameters. Eng Comput 38:1–14

Ku CY, Liu CY, Su Y et al (2018) Modeling of transient flow in unsaturated geomaterials for rainfall-induced landslides using a novel spacetime collocation method. Geofluids 2018:7892789

Ku CY, Hong LD, Liu CY et al (2021) Space–time polyharmonic radial polynomial basis functions for modeling saturated and unsaturated flows. Eng Comput. https://doi.org/10.1007/s00366-021-01519-z

Li X, Li XK, Wu YK et al (2022) Selection criteria of mesh size and time step in FEM analysis of highly nonlinear unsaturated seepage process. Comput Geotech 146:104712

List F, Radu FA (2016) A study on iterative methods for solving Richards' equation. Comput Geosci 20(2):341–353

Liu W (2017) A two-grid method for the semi-linear reaction–diffusion system of the solutes in the groundwater flow by finite volume element. Math Comput Simul 142:34–50

Liu CY, Ku CY, Xiao JE et al (2017) Numerical modeling of unsaturated layered soil for rainfall-induced shallow landslides. J Environ Eng Landsc Manag 25(4):329–341

Lott PA, Walker HF, Woodward CS et al (2012) An accelerated Picard method for nonlinear systems related to variably saturated flow. Adv Water Resour 38:92–101

Mendicino G, Senatore A, Spezzano G et al (2006) Three-dimensional unsaturated flow modeling using cellular automata. Water Resour Res 42(11):W11419

Mitra K, Pop IS (2019) A modified L-scheme to solve nonlinear diffusion problems. Comput Math Appl 77:1722–1738

Patankar SV (1980) Numerical heat transfer and fluid flow. CRC Press

Pachepsky Y, Timlin D, Rawls W (2003) Generalized Richards' equation to simulate water transport in unsaturated soils. J Hydrol 272(1–4):3–13

Parlange J Y, Hogarth W L, Barry D A, et al (1999) Analytical approximation to the solutions of Richards' equation with applications to infiltration, ponding, and time compression approximation. Adv Water Resour 23(2): 189–193

Pop IS, Schweizer B (2011) Regularization schemes for degenerate Richards equations and outflow conditions. Math Models Methods Appl Sci 21(8):1685–1712

Richards LA (1931) Capillary conduction of liquids through porous mediums. Physics 1(5):318–333

Seus D, Mitra K, Pop IS et al (2018) A linear domain decomposition method for partially saturated flow in porous media. Comput Methods Appl Mech Eng 333:331–355

Šimůnek J, Šejna M, Saito H, et al (2009) The HYDRUS-1D software package for simulating the one-dimensional movement of water, heat, and multiple solutes in variably-saturated media. Department of Environmental Sciences, University of California Riverside, Riverside, CA

Solin P, Kuraz M (2011) Solving the nonstationary Richards equation with adaptive hp-FEM. Adv Water Resour 34(9):1062–1081

Srivastava R, Yeh TCJ (1991) Analytical solutions for one-dimensional, transient infiltration toward the water table in homogeneous layered soils. Water Resour Res 27(5):753–762

Su NH (2014) Mass-time and space-time fractional partial differential equations of water movement in soils: theoretical framework and application to infiltration. J Hydrol 519:1792–1803

Tracy FT (2006) Clean two- and three-dimensional analytical solutions of Richards' equation for testing numerical solvers. Water Resour Res 42(8):8503

Wang HF, Anderson MP (1982) Introduction to groundwater modeling. Freeman

Wu L Z, Zhang L M, Li X (2016) One-dimensional coupled infiltration and de-formation in unsaturated soils subjected to varying rainfall. Int J Geomech 16(2): 06015004

Wang ZQ, Yin JF, Dou QY (2020) Preconditioned modified Hermitian and skew-Hermitian splitting iteration methods for fractional nonlinear Schrödinger equations. J Comput Appl Math 367:112420

Wu LZ, Huang RQ, Li X (2020a) Hydro-mechanical analysis of rainfall-induced landslides. Springer

Wu LZ, Zhu SR, Peng JB (2020b) Application of the Chebyshev spectral method to the simulation of groundwater flow and rainfall-induced landslides. Appl Math Model 80:408–425

Zeng JC, Zha YY, Yang JZ (2018) Switching the Richards' equation for modeling soil water movement under unfavorable conditions. J Hydrol 563:942–949

Zha Y, Yang J, Yin L et al (2017) A modified Picard iteration scheme for overcoming numerical difficulties of simulating infiltration into dry soil. J Hydrol 551:56–69

Zhang Z, Wang W, Yeh TJ et al (2016) Finite analytic method based on mixed-form Richards' equation for simulating water flow in vadose zone. J Hydrol 537:146–156

Zhu SR, Wu LZ, Shen ZH et al (2019) An improved iteration method for the numerical solution of groundwater flow in unsaturated soils. Comput Geotech 114:103113

Zhu SR, Wu LZ, Peng JB (2020) An improved Chebyshev semi-iterative method for simulating rainfall infiltration in unsaturated soils and its application to shallow landslides. J Hydrol 590:125157

Zhu SR, Wu LZ, Huang J (2022a) Application of an improved P(m)-SOR iteration method for flow in partially saturated soils. Comput Geosci 26:131–144

Zhu SR, Wu LZ, Song XL (2022b) An improved matrix split-iteration method for analyzing underground water flow. Eng Comput https://doi.org/10.1007/s00366-021-01551-z

Zhu SR, Wu LZ, Cheng P et al (2022c) Application of modified iterative method to simulate rainfall infiltration in unsaturated soils. Comput Geotech 148:104816

Zambra C E, Dumbser M, Toro E F, et al (2011) A novel numerical method of high-order accuracy for flow in unsaturated porous media. Int J Numer Meth Eng 89(2): 227–240

Chapter 4
Improved Linear and Nonlinear Iterative Methods for Rainfall Infiltration Simulation

4.1 Introduction

The linear infiltration equations obtained by discretizing Richards' equation need to be solved iteratively, including two approaches of linear and nonlinear iterations. The first method is to use numerical methods to directly numerically discretize Richards' equations to obtain nonlinear ordinary differential equations and then use nonlinear iterative methods to iteratively solve, such as Newton's method (Radu et al. 2006), Picard method (Lehmann and Ackerer 1998), and the *L*-method (List and Radu 2016). Among them, Paniconi and Putti (1994) found that Newton's method did not converge systematically, and sometimes the numerical results could be completely wrong. Casulli and Zanolli (2010) proposed a nested Newton method that can obtain quadratic convergence rates for arbitrary discrete time steps and all flow regimes. Brenner and Cancès (2017) introduced a new parameter to modify the Richards equation, which will be more robust when solving the equation using Newton's method. The Picard method can be considered as a simplified Newton method, which linearly converges. Since Newton's method increases algebraic complexity and computational cost for assembling the derivative terms in the Jacobian matrix (Zeng et al. 2018), the modified Picard method (Celia et al. 1990) has been widely studied for *h*-based and mixed forms of Richards' equation. However, since the coefficient matrix in each iteration needs to be re-evaluate in the Picard method, there is computationally expensive (Farthing and Ogden 2017). Some unfavorable numerical conditions such as water infiltration into dry and layered soils with very different hydraulic conductivities lead to numerical difficulties or even non-convergence in the conventional Picard method (Zha et al. 2017). Therefore, there has been increasing attention on the improvement of numerical discretization and computational efficiency of the Picard method (Celia et al. 1990; Zhu et al. 2020).

The second approach first needs to convert the Richards' equation into a linearized partial differential equation (Wu et al. 2020a, b), and then numerical methods are utilized to perform numerical discretization to obtain a set of linear equations and

© The Author(s) 2023
L. Wu and J. Zhou, *Rainfall Infiltration in Unsaturated Soil Slope Failure*,
SpringerBriefs in Applied Sciences and Technology,
https://doi.org/10.1007/978-981-19-9737-2_4

solve them iteratively. The classical linear iterative methods can efficiently solve systems of linear equations, such as the classic Jacobi iterative method (Hagemam and Young 1981), Gauss–Seidel iterative method (GS), and successive overrelaxation method (SOR). The GS method is an improvement on the Jacobi method, which usually takes less computation time, but it still requires more iterations to obtain a numerical solution than the SOR method. The Chebyshev semi-iterative method (CSIM) can improve the iterative convergence rate by introducing Chebyshev polynomials (Arioli and Scott 2014), but the computational cost of CSIM in each iteration is lower than that of traditional GS. Additionally, the Krylov subspace iterative method is also a common iterative method for solving linear equations, such as the conjugate gradient method, the generalized minimum residual method (GMRES), and the modified bi-conjugate gradient method (Dehghan and Mohammadi-Arani 2016). Theoretical analysis indicated that Hermitian/skew-Hermitian splitting iterative approaches (HSS) absolutely converge to a unique solution to a system of linear equations (Bai et al. 2010). HSS is an efficient linear iterative method for solving sparse non-Hermitian positive-definite equations (Dehghan and Shirilord 2019). Although HSS can converge unconditionally, it is time-consuming and impractical in real-world computation. Therefore, Bai et al. (2004) proposed the inexact Hermitian/skew-Hermitian splitting iterative approaches (IHSS) combined with the Krylov subspace iterative method to improve computational efficiency. However, IHSS is still time-consuming to solve linear equations for finer mesh discretization. Due to some shortcomings of conventional iterative methods, it is of great significance to improve numerical stability, accuracy, and convergence rate (Lott et al. 2012; Deng and Wang 2017; Mitra and Pop 2019; Illiano et al. 2021; Su et al. 2022).

The improvement methods for common accelerated convergence include extrapolation, error correction, Chebyshev polynomial acceleration (Arioli and Scott 2014), and preconditioning technique (Benzi 2002). For example, Lott et al. (2012) studied the effectiveness of the improved Picard method by the Anderson acceleration method for the variable saturated flow problem. Arioli and Scott (2014) analyzed an improved iterative method of Chebyshev polynomials that can be applied to accelerate iterative optimization without losing numerical stability. Wang and Zhang (2003) proposed a class of parallel multi-step sequential preprocessing strategies, which can significantly improve the computational efficiency and stability of solving linear equations. Briggs et al. (2000) proposed a multi-grid correction method to quickly eliminate the iterative error in the iterative process, thereby obtaining a faster convergence rate. Recently, preconditioning methods can effectively reduce the condition number of the iterative matrix of linear equations, thereby improving the computational convergence rate (Benzi 2002; Zhu et al. 2022a, b, c), including the left preconditioning, right preconditioning, and two-side preconditioning (Liu et al. 2015). The preconditioning technique converts the original system of linear equations into a system that is easier to solve, thereby improving the convergence rate of the iterative process (Benzi 2002).

With the higher accuracy of the numerical solutions to solve the infiltration problem, the scale of the system of algebraic equations increases, and then the calculation time also increases. Therefore, the numerical methods and iterative methods are constantly improving and innovating. The research based on traditional methods and newly developed and new algorithms is of great significance for the cross-development of geotechnical engineering, geological engineering, and other majors. In this chapter, the Richards' equation can be transformed into a linearized Richards' equation using an exponential function. Furthermore, the numerical method is used to discretize the linear RE, and the linear and nonlinear iterative methods are employed to evaluate computational efficiency and accuracy using different iterative methods.

4.2 A Chebyshev Semi-iterative Method with Preconditioner

For the system of linear equations $\mathbf{Ah} = \mathbf{b}$ obtained by numerically discreting Richards' equation, the left preprocessing method is given by:

$$\mathbf{M}^{-1}\mathbf{Ah} = \mathbf{M}^{-1}\mathbf{b} \tag{4.1}$$

Consequently, the condition number of $\mathbf{M}^{-1}\mathbf{A}$ is greatly reduced.

Let $\mathbf{A}$ be a non-singular matrix. Matrix $\mathbf{A}$ can be split into $\mathbf{A} = \mathbf{D} - \mathbf{Q}$, with $\mathbf{D}$ a non-singular matrix and is the diagonal matrix of matrix $\mathbf{A}$. Then, if $\mathbf{H} = \mathbf{D}^{-1}\mathbf{Q}$ has spectral radius less than 1.0, one obtains:

$$\mathbf{A}^{-1} = \left(\sum_{k=0}^{\infty} \mathbf{H}^k\right)\mathbf{D}^{-1} \tag{4.2}$$

Matrix $\mathbf{M}$ is defined as:

$$\mathbf{M} = \mathbf{D}\left(\mathbf{I} + \mathbf{H} + \cdots + \mathbf{H}^{m-1}\right)^{-1} \tag{4.3}$$

where m is a natural number. Furthermore, matrix $\mathbf{M}$ is considered as a preconditioner to estimate the condition number of $\mathbf{M}^{-1}\mathbf{A}$. From Eq. (4.3) and $\mathbf{A} = \mathbf{D} - \mathbf{Q}$, one can obtain:

$$\mathbf{M}^{-1}\mathbf{A} = \left(\mathbf{I} + \mathbf{H} + \cdots + \mathbf{H}^{m-1}\right)\mathbf{D}^{-1}(\mathbf{D} - \mathbf{Q}) = \mathbf{I} - \mathbf{H}^m \tag{4.4}$$

If $\lambda_1, \ldots, \lambda_n$ are the eigenvalue of matrix $\mathbf{H}$, the eigenvalues of $\mathbf{M}$ are $1 - \lambda_i^m$. If $\mathbf{D}$ and $\mathbf{A}$ are symmetric positive-definite matrices, then the eigenvalues of $\mathbf{H}$ are real. Assuming that $\lambda_1 \leq \cdots \leq \lambda_n < 1$, then one can have:

$$\text{cond}\left(\mathbf{M}^{-1}\mathbf{A}\right) = \frac{\lambda_{\max}\left(\mathbf{M}^{-1}\mathbf{A}\right)}{\lambda_{\min}\left(\mathbf{M}^{-1}\mathbf{A}\right)} = \begin{cases} \frac{1-\lambda_1^m}{1-\lambda_n^m}, & \text{if } \lambda_1 \geq 0 \text{ or } \lambda_1 < 0, \ m \text{ is odd,} \\ \frac{1-\sigma^m}{1-\lambda_n^m}, & \text{if } \lambda_1 < 0, \ |\lambda_n| \geq |\lambda_1|, \ m \text{ is even,} \\ \frac{1-\sigma^m}{1-\lambda_1^m}, & \text{if } \lambda_1 < 0, \ |\lambda_1| > |\lambda_n|, \ m \text{ is even,} \end{cases}$$

$$(4.5)$$

where $\sigma = \min_i |\lambda_i|$.

From Eq. (4.5), the condition number of matrix $\mathbf{M}^{-1}\mathbf{A}$ decreases with increasing m. Hence, the preprocessing can effectively decrease the condition number of this matrix and enhance the convergence rate.

Based on the Chebyshev polynomial acceleration process and the Gauss–Seidel iterative method of preconditioning, this chapter develops an improved preconditioning Chebyshev semi-iterative method (P-CSIM). First, the improved Gauss–Seidel iterative format using the preconditioning (Eq. 4.3) is written as:

$$\mathbf{h}^{k+1} = \mathbf{G}_{\mathbf{M}}\mathbf{h}^k + \mathbf{b}_{\mathbf{M}} \qquad (4.6)$$

where $\mathbf{G}_{\mathbf{M}}$ denotes an iterative matrix and $\mathbf{s}_{\mathbf{M}}$ denotes a column matrix of preprocessing. Furthermore, the error vector for the general polynomial acceleration process of Eq. (4.6) is expressed as:

$$\varepsilon^k = Q_k(\mathbf{G}_{\mathbf{M}})\varepsilon^0 \qquad (4.7)$$

where $Q_k(\mathbf{G}_{\mathbf{M}}) \equiv a_{k,0}\mathbf{I} + a_{k,1}\mathbf{G}_{\mathbf{M}} + \cdots + a_{k,k}\mathbf{G}_{\mathbf{M}}^k$ is a matrix polynomial, and the only additional condition being $\sum_{i=0}^k a_{k,i} = 1$. The virtual spectral radius of the matrix $Q_k(\mathbf{G}_{\mathbf{M}})$ is then defined as:

$$\overline{\mathbf{S}}(Q_k(\mathbf{G}_{\mathbf{M}})) = \max_{m(\mathbf{G}_{\mathbf{M}}) \leq x \leq M(\mathbf{G}_{\mathbf{M}})} |Q_k(x)| \qquad (4.8)$$

where $M(\mathbf{G}_{\mathbf{M}})$ and $m(\mathbf{G}_{\mathbf{M}})$ denote the algebraic maximum and minimum eigenvalues of $\mathbf{G}_{\mathbf{M}}$, respectively.

It can be proved that the matrix polynomial $Q_k(\mathbf{G}_{\mathbf{M}})$ that ensures $\overline{\mathbf{S}}(Q_k(\mathbf{G}_{\mathbf{M}}))$ reaches a minimum is unique and can be defined by a Chebyshev polynomial. Furthermore, it is needed to find the polynomial $P_k(x)$, for which $P_k(1) = 1$, and satisfies the following conditions (Theorem 4.2.1 of Hagemam and Young 1981):

$$\max_{m(\mathbf{G}_{\mathbf{M}}) \leq x \leq M(\mathbf{G}_{\mathbf{M}})} |P_k(x)| \leq \max_{m(\mathbf{G}_{\mathbf{M}}) \leq x \leq M(\mathbf{G}_{\mathbf{M}})} |Q_k(x)| \qquad (4.9)$$

where $Q_k(x)$ satisfies $Q_k(1) = 1$. Assume that the eigenvalues of $\mathbf{G}_{\mathbf{M}}$ are real numbers, $m(\mathbf{G}_{\mathbf{M}}) = \alpha$, and $M(\mathbf{G}_{\mathbf{M}}) = \beta$. According to Theorem 2.2.1 of Hagemam and Young (1981, p. 21), let:

$$\alpha = \lambda_1 \leq \lambda_2 \leq \cdots \leq \lambda_n = \beta < 1, \quad \alpha \neq \beta \qquad (4.10)$$

To obtain $P_k(x)$, a linear transformation is required as follows:

$$r = \frac{2x - \alpha - \beta}{\beta - \alpha} \tag{4.11}$$

Moreover, one can obtain:

$$\begin{cases} r\big|_{x=\alpha} = -1, \quad r\big|_{x=\beta} = 1, \\ x = \frac{(\beta-\alpha)r+(\beta+\alpha)}{2}, \\ Q_k(x) = Q_k\left(\frac{(\beta-\alpha)r+(\beta+\alpha)}{2}\right) =: P_k(r). \end{cases} \tag{4.12}$$

A new parameter z^* is introduced as follows:

$$z^* = r\big|_{x=1} = \frac{2 - \alpha - \beta}{\beta - \alpha} \tag{4.13}$$

Note that $z^* > 1$ and $P_k(z^*) = Q_k(1) = 1$. Here, $P_k(x)$ can be defined as:

$$P_k(x) \equiv T_k(r)/T_k(z^*) \tag{4.14}$$

Additionally, the kth Chebyshev polynomials on $T_k(r)$ for any non-negative integer k and r can be defined by the following 3-term recurrence equation (Arioli and Scott 2014):

$$\begin{cases} T_0(r) = 1, \quad\quad\quad\quad\quad T_1(r) = r, \\ T_{k+1}(r) = 2r T_k(r) - T_{k-1}(r), \ k \geq 1. \end{cases} \tag{4.15}$$

Using Theorem 4.2.1 (Hagemam and Young 1981), it is easy to prove that polynomials $P_k(x)$ satisfy the recurrence relation. Therefore, according to Eqs. (4.6), (4.14), and (4.15) and Theorem 3.2.1 (Hagemam and Young 1981, p. 41), the 3-term recurrence equations for the P-CSIM are written as:

$$\begin{cases} \mathbf{h}^1 = \overline{\gamma}(\mathbf{G}_M\mathbf{h}^0 + \mathbf{b}_M) + (1 - \overline{\gamma})\mathbf{h}^0, \\ \mathbf{h}^{k+1} = \overline{\rho}_{k+1}\left[\overline{\gamma}(\mathbf{G}_M\mathbf{h}^k + \mathbf{b}_M) + (1 - \overline{\gamma})\mathbf{h}^k\right] + \left(1 - \overline{\rho}_{k+1}\right)\mathbf{h}^{k-1} \ k \geq 1, \end{cases} \tag{4.16}$$

in which

$$\overline{\gamma} = \frac{2}{2 - \beta - \alpha} \tag{4.17}$$

$$\overline{\rho}_{k+1} = \frac{2r(1)T_k(r(1))}{T_{k+1}(r(1))} \tag{4.18}$$

Equation (4.16) can be further expressed as:

$$\begin{cases} \mathbf{h}^1 = \frac{2}{2-\beta-\alpha}\left(\mathbf{G}_M\mathbf{h}^0 + \mathbf{b}_M\right) - \frac{\alpha+\beta}{2-\beta-\alpha}\mathbf{h}^0, \\ \mathbf{h}^{k+1} = 2\frac{2\mathbf{G}_M-(\alpha+\beta)\mathbf{I}}{\beta-\alpha}\frac{T_k(z^*)}{T_{k+1}(z^*)}\mathbf{h}^k - \frac{T_{k-1}(z^*)}{T_{k+1}(z^*)}\mathbf{h}^{k-1} + \frac{4\mathbf{b}_M}{\beta-\alpha}\frac{T_k(z^*)}{T_{k+1}(z^*)} \quad k \geq 1, \end{cases} \tag{4.19}$$

where $T_k(z^*)$ is recursively calculated using Eq. (4.15); here, $\mathbf{I}$ is an identity matrix.

From Eq. (4.19), the convergence rate of P-CSIM mainly depends on the accuracy in estimating the maximum and minimum eigenvalues β and α of the iterative matrix $\mathbf{G}_M$. This chapter assumes $0 < -\alpha = \beta < 1$ based on the basic assumptions of Hagemam and Young (1981). By the above processing, the proposed P-CSIM achieves a better convergence and computational efficiency in solving the linear algebraic equations.

A brief implementation flowchart of P-CSIM is described in Fig. 4.1 (Zhu et al. 2020). To compare the speed-up of the improved method, the speed-up ratio is defined as:

$$S_{GS/P} = \frac{T_{GS}}{T_{P\text{-CSIM}}} \tag{4.20}$$

where T_{GSIM} and $T_{P\text{-CSIM}}$ represent the computing times for GSIM and P-CSIM.

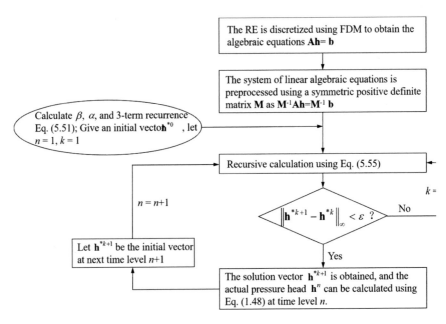

Fig. 4.1 Brief flowchart of P-CSIM

4.2.1 Numerical Tests

The mathematical model, boundary conditions, and unsaturated soil parameters of the numerical tests are consistent with Chap. 3, and the analytical solutions are expressed as Eqs. (2.3)–(2.5). The total duration of the simulation was 5 h, and the convergence criterion was set to 10^{-8} for the different iteration schemes. In preprocessing, the expansion order m is set to 20. To verify the accuracy, efficiency, and robustness of the proposed method, the time steps were set to 0.01 h, 0.005 h, 0.002 h, and 0.001 h, respectively, and the grid sizes were set to 0.4 m, 0.2 m, and 0.1 m.

Figure 4.2 demonstrates the relationship between expansion order m and the condition number for different time steps in this test. The condition number of coefficient matrix A increases with the time step. However, the condition number can be decreased to 1.0 through the preprocessing method. Additionally, Fig. 4.2a–d indicates that the condition number is close to 1.0 when $m = 6$, indicating that preprocessing effectively improves the ill-conditioning of the matrix.

Table 4.1 sums up the iterations, the computing time obtained from the different iterative schemes with different grid sizes, and the time steps in this test. Compared with other methods, the improved P-CSIM has the lowest iterations and the least computational cost at different time steps and grid sizes. When $\Delta z = 0.1$, the

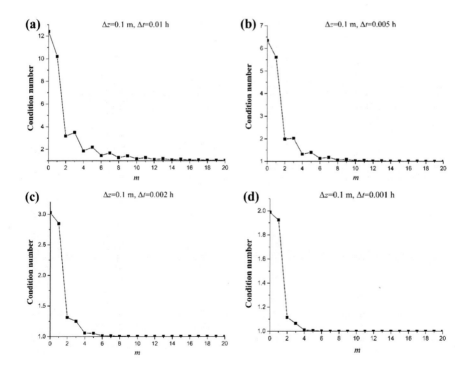

Fig. 4.2 Relationship between parameter m and condition number at different time steps

iterations of the P-CSIM are much smaller than that of other methods (Fig. 4.3a). Moreover, the speed-up ratio of P-CSIM relative to the other three methods at $\Delta t = 0.005$ (Fig. 4.3b) shows that it is the largest with JIM, CSIM, and GSIM following in sequence. This result also demonstrates that the proposed P-CSIM represents a vast improvement over CSIM.

RMSE reaches an order of 10^{-5} at $t = 5$ h (Fig. 4.4) and decreases with decreasing grid size for different iterative methods. Moreover, the accuracy of the proposed P-CSIM shows a slight improvement compared with that of GSIM and CSIM.

The results illustrate that the improved P-CSIM has better computational efficiency, stability, and accuracy than the other three iterative methods.

Table 4.1 Numerical results of different iterative methods (Zhu et al. 2020)

Conditions		Iterations				Computing time (s)			
Δz (m)	Δt (h)	JIM	GSIM	CSIM	P-CSIM	JIM	GSIM	CSIM	P-CSIM
0.4	0.01	13	10	10	2	0.0885	0.0471	0.0522	0.0203
	0.005	10	8	8	2	0.1461	0.0734	0.0790	0.0345
	0.002	8	7	7	2	0.2650	0.1523	0.1647	0.0669
	0.001	7	6	6	2	0.4482	0.2632	0.2899	0.1371
0.2	0.01	29	19	17	3	0.2392	0.1047	0.1569	0.0562
	0.005	19	14	13	2	0.3184	0.1515	0.2388	0.0573
	0.002	12	10	9	2	0.4881	0.3156	0.3681	0.1137
	0.001	9	8	8	2	0.7564	0.4459	0.6684	0.2321
0.1	0.01	85	49	39	5	1.1104	0.5001	1.0043	0.1409
	0.005	49	29	24	4	1.2838	0.6354	1.2199	0.2570
	0.002	25	17	16	3	1.6605	0.9149	2.0057	0.3988
	0.001	17	12	12	2	2.2653	1.2625	2.9351	0.5303

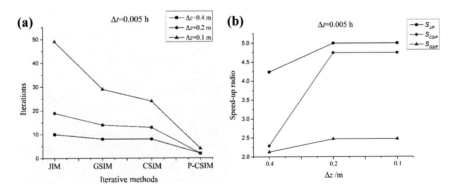

Fig. 4.3 Numerical results of different iterative methods at $t = 0.005$ h: **a** iterations and **b** speed-up ratio

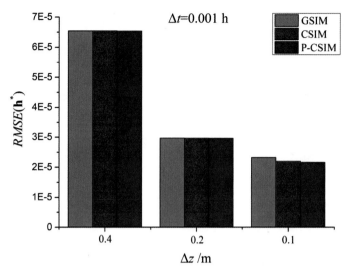

Fig. 4.4 Root mean squared error (RMSE) associated with several grid sizes at $t = 5$ h for the different iterative methods

4.3 Improved Gauss–Seidel Method

When the first-order linear stationary iterative methods (Eq. 4.6) or other iterative schemes to solve linear equations are used, the iterative process may or may not converge. Even if it does, it may converge very slowly. In either condition, this chapter hopes to find an improved method to make non-convergent formats converge, and slow-convergent formats converge faster. The commonly used acceleration methods include extrapolation method, integral correction method, Chebyshev polynomial acceleration method, and preconditioning method.

4.3.1 Integral Correction Method

The integral correction method is similar to Anderson acceleration method (Walker and Ni 2011; Both et al. 2018). Firstly, the mutually different approximate solutions $\mathbf{h}_1$, $\mathbf{h}_2$, ... $\mathbf{h}_c$ $(c > 1)$ can be obtained according to the basic iterative method, and $\mathbf{Ah}_i \neq \mathbf{b}$ $(i = 1, 2, ..., c)$. Then the information provided by $\mathbf{h}_i$ can be used to construct the vector $\mathbf{h}$ so that it is closer to the exact solution of the linear equations than $\mathbf{h}_i$, that is, there is:

$$\|\mathbf{b} - \mathbf{Ah}\|_2 < \min_{1 \leq i \leq c} \|\mathbf{b} - \mathbf{Ah}_i\|_2 \tag{4.21}$$

The vector $\mathbf{h}$ that satisfies Eq. (4.21) is the correction solution of the linear equations with respect to $\mathbf{h}_1$, $\mathbf{h}_2$, … $\mathbf{h}_c$. The process of determining $\mathbf{h}$ is the correction process, then the integral correction model can be expressed as:

$$\begin{cases} \mathbf{h} = \sum_{i=1}^{c} \beta_i \mathbf{h}_i, \quad \sum_{i=1}^{c} \beta_i = 1, \\ \|\mathbf{b} - \mathbf{Ah}\|_2 = \min \end{cases} \quad (4.22)$$

Select arbitrarily $s \in \{1, 2, \ldots, c\}$, then

$$\mathbf{h} = \mathbf{h}_s + \sum_{i \neq s} \beta_i (\mathbf{h}_i - \mathbf{h}_s),$$

$$\mathbf{r}(\mathbf{h}) = \mathbf{b} - \mathbf{Ah} = (\mathbf{b} - \mathbf{Ah}_s) - \sum_{i \neq s} \beta_i (\mathbf{Ah}_i - \mathbf{Ah}_s) = \mathbf{r}_s + \sum_{i \neq s} \beta_i (\mathbf{r}_i - \mathbf{r}_s)$$

$$(4.23)$$

Denote

$$\delta_i = \mathbf{r}_i - \mathbf{r}_s, \quad \mathbf{Q}_s = \left[\delta_1, \ldots, \delta_{s-1}, \delta_{s+1}, \ldots, \delta_c \right],$$
$$\mathbf{y}_s = (\beta_1, \ldots, \beta_{s-1}, \beta_{s+1}, \ldots, \beta_c)^{\mathrm{T}}$$

Therefore, Eq. (4.23) can be rewritten as $\mathbf{r}(\mathbf{h}) = \mathbf{r}_s + \mathbf{Q}_s \mathbf{y}_s$, so that,

$$\mathbf{h} = \sum_{i=1}^{c} \beta_i \mathbf{h}_i = \beta_1 \mathbf{h}_1 + \beta_2 \mathbf{h}_2 + \cdots + \beta_c \mathbf{h}_c \quad (4.24)$$

makes $\|\mathbf{r}(\mathbf{h})\|_2 = \min$, which is equivalent to $\mathbf{y}_s \in \mathbb{C}^{c-1}$ makes $\|\mathbf{r}_s + \mathbf{Q}_s \mathbf{y}_s\|_2 = \min$, and its minimal norm solution can be expressed as:

$$\mathbf{y}_s = -\mathbf{Q}_s^{\dagger} \mathbf{r}_s \quad (4.25)$$

where $\mathbf{Q}_s^{\dagger}$ is the generalized inverse matrix of $\mathbf{Q}_s$. The parameter β_s can be obtained by $\beta_s = 1 - \sum_{i \neq s} \beta_i$, and then the correction solution $\mathbf{h}$ can be determined using Eq. (4.24). In particular, for $c = 2$, one can gain:

$$\begin{cases} \beta_1 = y_2 = -\mathbf{Q}_2^{\dagger} \mathbf{r}_2 = -\delta_1^{\dagger} \mathbf{r}_2 = -\frac{(\mathbf{r}_1 - \mathbf{r}_2)^H \mathbf{r}_2}{\|\mathbf{r}_1 - \mathbf{r}_2\|_2^2} \\ \beta_2 = 1 - \beta_1 = \frac{(\mathbf{r}_1 - \mathbf{r}_2)^H \mathbf{r}_1}{\|\mathbf{r}_1 - \mathbf{r}_2\|_2^2} \end{cases} \quad (s = 2) \quad (4.26)$$

$$\begin{cases} \beta_2 = y_1 = -\mathbf{Q}_1^{\dagger} \mathbf{r}_1 = -\delta_2^{\dagger} \mathbf{r}_1 = -\frac{(\mathbf{r}_2 - \mathbf{r}_1)^H \mathbf{r}_1}{\|\mathbf{r}_2 - \mathbf{r}_1\|_2^2} \\ \beta_1 = 1 - \beta_2 = \frac{(\mathbf{r}_2 - \mathbf{r}_1)^H \mathbf{r}_2}{\|\mathbf{r}_2 - \mathbf{r}_1\|_2^2} \end{cases} \quad (s = 1) \quad (4.27)$$

It can be seen that the results of Eqs. (4.26) and (4.27) are consistent. Therefore, the improved Gauss–Seidel iterative method (IC(c)-GS) based on the integral correction method can be summarized as follows:

(1) Input the matrix $\mathbf{A}$, the right vector $\mathbf{b}$, the initial vector $\mathbf{h}_0^*$, the integer c, and $s = c$, and let $k = 1$.
(2) Calculate $\mathbf{h}_i^{*k}$ using Gauss–Seidel iterative method, $i = 1, 2, \ldots, c$.
(3) Calculate $\mathbf{y}_s^k$ and β_s^k by Eq. (4.25).
(4) Calculate $\mathbf{h}^{*k} = \beta_1^k \mathbf{h}_1^{*k} + \beta_2^k \mathbf{h}_2^{*k} + \cdots + \beta_c^k \mathbf{h}_c^{*k}$.
(5) If the accuracy meets the requirements, the calculation stops and $\mathbf{h}^{*k}$ is output as the approximate solution of the linear equations. Otherwise, let $k = k + 1$ and return to step (2) to continue the next iteration.

4.3.2 Multistep Preconditioner Method

For a given system of linear equations, $\mathbf{A}\mathbf{h}^* = \mathbf{b}$, the condition number may be used to evaluate whether a given non-singular matrix $\mathbf{A}$ is ill-conditioned. The condition number for a non-singular matrix $\mathbf{A}$ can be defined as:

$$\text{Cond}(\mathbf{A}) = \|\mathbf{A}\| \cdot \|\mathbf{A}\|^{-1} \tag{4.28}$$

where the matrix norm is the Frobenius norm.

The preconditioning method can be an acceleration technique for the iterative solution of the system of linear equations. It can greatly improve the ill-condition of the original system of linear equations, thereby accelerating the convergence rate of the iterative method (Benzi 2002). The preconditioning process is usually to find the matrix $\mathbf{M}$ (preconditioner) for the linear equations $\mathbf{A}\mathbf{h}^* = \mathbf{b}$. There are many choices and forms of $\mathbf{M}$. The preconditioning on the left for $\mathbf{A}\mathbf{h}^* = \mathbf{b}$ can be written as:

$$\mathbf{M}^{-1}\mathbf{A}\mathbf{h}^* = \mathbf{M}^{-1}\mathbf{b} \tag{4.29}$$

$\mathbf{A}\mathbf{h}^* = \mathbf{b}$ is preconditioned from the right:

$$\mathbf{A}\mathbf{M}^{-1}\mathbf{y} = \mathbf{b}, \quad \mathbf{h}^* = \mathbf{M}^{-1}\mathbf{y} \tag{4.30}$$

The preconditioning on both sides of $\mathbf{A}\mathbf{h}^* = \mathbf{b}$ is:

$$\mathbf{M}_1^{-1}\mathbf{A}\mathbf{M}_2^{-1}\mathbf{z} = \mathbf{M}_1^{-1}\mathbf{b}, \quad \mathbf{h}^* = \mathbf{M}_2^{-1}\mathbf{z}, \quad \mathbf{M} = \mathbf{M}_1\mathbf{M}_2 \tag{4.31}$$

Based on the preconditioner increasing the complexity of the iterative method, here the preconditioner should be simple to construct and greatly improve the convergence rate of the iterative method. Therefore, the left preconditioning (Eq. 4.29) is adopted in this chapter. A symmetric Gauss–Seidel (SGS) preconditioner is written as:

$$\mathbf{M}_{SGS} = (\mathbf{D} - \mathbf{L})\mathbf{D}^{-1}(\mathbf{D} - \mathbf{U}) \qquad (4.32)$$

It is easy to know that combining the split matrix of the GS can quickly construct the preconditioner (Eq. 4.32). This preconditioning technique has almost no construction cost and is easy to implement. However, the result of one-step preconditioning is sometimes not good enough. Furthermore, a multistep SGS preconditioner is proposed to further obtain a sufficiently good and easier-to-solve system of the linear equations:

$$\prod_{i=1}^{m} \mathbf{M}_{SGS}^{-1}{}^{i}\mathbf{Ah}^* = \prod_{i=1}^{m} \mathbf{M}_{SGS}^{-1}{}^{i}\mathbf{b} \qquad (4.33)$$

where m represents the number of steps of the multistep preconditioner. Generally, a multistep preconditioner with a small number of steps that can lead to good convergence results in the iterative method, and a multistep preconditioner with a large number of steps may be more robust, but it may also cause higher computational cost of the iterative method. The empirical number of steps should be 2, 3, or 4, which can provide the preconditioning iterative method with good convergence rate. The Gauss–Seidel iterative method with multistep preconditioner (MP(m)-GS) can be summarized as follows:

(1) Give the matrix $\mathbf{A}$, the right vector $\mathbf{b}$, the initial vector $\mathbf{h}_0^*$, the filter parameter τ, and the number of steps m, and let $\mathbf{A}_1 = \mathbf{A}$, $k = 1$.

(2) Circularly calculate $\mathbf{A}_{i+1}$:
     For ($i = 1$; $i \leq m$; i++).
     Compute an SGS preconditioner $\mathbf{M}_{SGS}^{-1}{}^{i}$ using $\mathbf{A}_i$.
     Drop the small entries in the matrix $\mathbf{M}_{SGS}^{-1}{}^{i}$ relative to the parameter τ.
     Compute $\mathbf{A}_{i+1} = \mathbf{M}_{SGS}^{-1}{}^{i}\mathbf{A}_i$.
     End the cycle.

(3) $\prod_{i=1}^{m} \mathbf{M}_{SGS}^{-1}{}^{i}$ is used as a preconditioner to perform left preconditioning on $\mathbf{Ah}^* = \mathbf{b}$.

(4) The GS is used to solve the preconditioned linear equations (Eq. 4.33).

(5) If the accuracy is satisfied, the calculation stops and $\mathbf{h}^{*k}$ is output as the approximate solution of the linear equations. Otherwise, let $k = k + 1$ and return to step (3) to continue the next iteration.

The parameter τ can ensure the sparsity of the preconditioning matrix, thereby reducing the computational cost of the preconditioning process (Wang and Zhang 2003). In this chapter, the parameter τ is uniformly set to 10^{-4}. It can be easily found that when $m = 0$, MP(m)-GS completely degenerates to GS.

4.3.3 Mixed Method

The integral correction method utilizes the solution vector error correction to make the iterative method obtain a faster convergence rate, while the multistep preconditioning method preprocesses the original system of linear equations into a more easily solved system of linear equations, making the subsequent iterative method more efficient. The two methods are different. Therefore, the two methods are combined to propose an improved Gauss–Seidel iterative method (ICMP(m)-GS) with multistep preconditioner based on the integral correction method. The method is summarized as follows:

(1) Give the matrix $\mathbf{A}$, the right vector $\mathbf{b}$, the initial vector $\mathbf{h}_0^*$, the integer c, the filter parameter τ, and the number of steps m, and let $k = 1$.
(2) Calculate $\mathbf{M}_{SGS}^{-1}{}^i$ ($i = 1, 2, ..., m$) cyclically, and drop the small entries in the matrix $\mathbf{M}_{SGS}^{-1}{}^i$ relative to the parameter τ.
(3) $\prod_{i=1}^{m} \mathbf{M}_{SGS}^{-1}{}^i$ is used as a preconditioner to perform left preconditioning on $\mathbf{A}\mathbf{h}^* = \mathbf{b}$, and GS is used to calculate $\mathbf{h}_i^{*k}$ ($i = 1, 2, ..., c$).
(4) Calculate $\mathbf{y}_s^k$ and β_s^k according to Eq. (4.25), and calculate $\mathbf{h}^{*k} = \beta_1^k \mathbf{h}_1^{*k} + \beta_2^k \mathbf{h}_2^{*k} + \cdots + \beta_c^k \mathbf{h}_c^{*k}$.
(5) If $\mathbf{h}^{*k}$ converges according to the convergence criterion, then stop; otherwise, go to step (3).

To verify the calculation efficiency, the speed-up ratio is defined as:

$$S_{GS/MP(m)\text{-}GS} = \frac{T_{GS}}{T_{MP(m)\text{-}GS}} \tag{4.34}$$

where T_{GS} and $T_{MP(m)\text{-}GS}$ are the runtimes of GS and MP(m)-GS, respectively.

4.3.4 Numerical Tests

4.3.4.1 1D Steady-State Unsaturated Infiltration

The mathematical model is shown in Fig. 1.1. The model parameters are: $\alpha = 8 \times 10^{-3}$, $\theta_s = 0.35$, $\theta_r = 0.14$, and $h_d = -10^3$ m. For the integral correction method, to test the influence of the parameter c on the improved iterative method, c is set to 2, 5, 10, and 20. For the multistep preconditioning method, m is 1, 2, 3, and 4 to verify the effectiveness of the algorithm. In Fig. 4.5, the change in the space step Δz has a small impact on the acceleration effect of the IC(c)-GS iterative method, while the change of the parameter c has a greater impact on it. When the parameter c is less than 10, the speed-up ratio of IC(c)-GS relative to the GS is relatively small. And when $c \geq 10$, the speed-up ratio increases greatly. It can be found that the acceleration effects of $c = 10$ and $c = 20$ are roughly similar. Therefore, the parameter c of the IC(c)-GS is set to 10 in this chapter. In Fig. 4.6, the convergence rate of the Jacobi is the slowest,

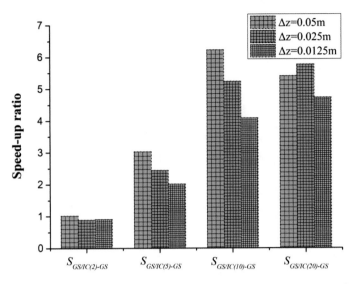

Fig. 4.5 Comparison of the speed-up ratio of IC(c)-GS relative to GS under different parameter c

followed by the GS, and the SOR is faster than the Jacobi and GS. It is easy to obtain that when $c \geq 10$, the convergence rate of IC(c)-GS is faster than SOR. When $m = 3$ and 4, the convergence rate of MP(m)-GS is also faster than SOR. For the proposed mixed method ICMP(m)-GS, it can be found from the enlarged convergence graph that its convergence rate is faster than SOR, IC(c)-GS, and MP(m)-GS, regardless of the value of m, and increases with the increase of m.

In Fig. 4.7, the speed-up ratio of MP(m)-GS and ICMP(m)-GS relative to GS is very large when $\Delta z = 0.025$ m, and the acceleration effect is much better than IC(c)-GS. The acceleration effect of ICMP(m)-GS is better than MP(m)-GS. Additionally, the speed-up ratio of ICMP(m)-GS relative to GS increases slowly when the parameter m is 3 and 4, which is related to the increase of the convergence rate. In other words, the improved method ICMP(m)-GS increases the computational complexity and cost as increasing m, but the number of iterations does not greatly decrease. The numerical results are listed in Table 4.2. The number of iterations and running time of GS and ICMP(3)-GS increase as Δz decreases, but the number of iterations of ICMP(3)-GS is much smaller than that of GS, and the difference is about 10,000 times. The running time of ICMP(3)-GS is much shorter than that of GS, and the difference is about 100 times.

4.3.4.2   1D Transient Unsaturated Infiltration

The boundary conditions can be written as follows:

$$h(z = 0) = h_\mathrm{d} \qquad\qquad (4.35)$$

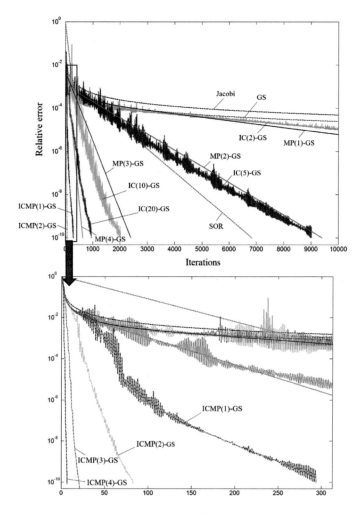

Fig. 4.6 Comparison of convergence rates for the different iterative methods

$$h(z = 10) = 0 \tag{4.36}$$

The parameters are set to: $\alpha = 1 \times 10^{-4}$, $\theta_s = 0.50$, $\theta_r = 0.11$, and the saturated permeability coefficient $k_s = 2.5 \times 10^{-8}$ m/s. The total simulation time is 10 h. To verify the calculation efficiency of the proposed method, the space step is taken as 0.05 m, 0.025 m, and 0.0125 m, and the time step is 0.2 h, 0.1 h, and 0.05 h, respectively.

Figure 4.8 shows the changes in the condition number of the coefficient matrix **A** with and without the algorithm MP(m) under different numerical discrete conditions. The condition number of the coefficient matrix increases with the decrease of the

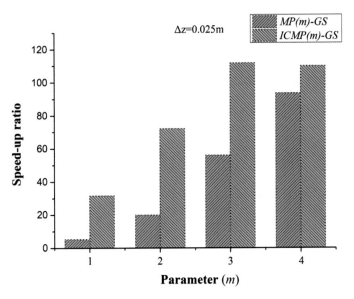

Fig. 4.7 Effect of parameter m on the improved iterative methods MP(m)-GS and ICMP(m)-GS for $\Delta z = 0.025$ m

Table 4.2 Numerical solutions of steady-state infiltration

Conditions	Number of iterations		Running time (s)	
Δz (m)	GS	ICMP(3)-GS	GS	ICMP(3)-GS
0.0500	57,942	8	2.43	0.06
0.0250	209,201	21	28.09	0.25
0.0125	746,728	81	536.99	2.03

space step and decreases with the decrease of the time step. When the preconditioning method is not used, the condition number reaches in the order of 10^4. However, the condition number can be greatly reduced using the algorithm MP(m) and decrease with the increase of the parameter m. Particularly after using algorithm MP(4), the condition number even approaches 1.0, which indicates that the multistep preconditioning process can effectively improve the ill-condition of the system of the linear equations.

Tables 4.3 and 4.4 list the average number of iterations and running time of the conventional iterative methods GS and SOR and the improved method ICMP(m)-GS. First of all, it can be seen that the performance of SOR is better than that of GS. However, the number of iterations of SOR is not stable under a smaller space step. Compared with GS and SOR, the number of iterations and running time of the improved ICMP(m)-GS is less than that of the conventional method. The number of iterations increases with the decrease of Δz and decreases with the decrease of Δt. The running time increases as Δz and Δt decrease. At the same time, the

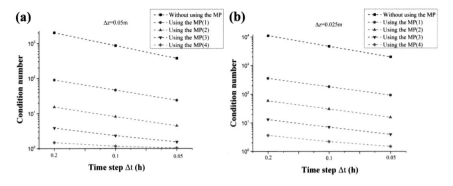

Fig. 4.8 Condition number with and without the MP algorithm for different conditions **a** $\Delta z = 0.05$ m and **b** $\Delta z = 0.025$ m

Table 4.3 Number of iterations for solving steady-state infiltration

Conditions		Average number of iterations for each time step					
Δz (m)	Δt (h)	GS	SOR	ICMP(1)-GS	ICMP(2)-GS	ICMP(3)-GS	ICMP(4)-GS
0.05	0.2	3071	274	13	6	3	2
	0.1	1639	222	9	5	3	2
	0.05	868	160	7	4	3	2
0.025	0.2	11,262	545	32	11	5	3
	0.1	6019	401	20	8	4	3
	0.05	3185	403	14	6	4	2

performance of ICMP(m)-GS with $m = 3$ and 4 is much better than that of GS and SOR, and it has a very significant acceleration effect. Besides, the acceleration effect of the improved method ICMP(m)-GS also has certain rules under different grid sizes. In Fig. 4.9, with the decrease of the space step Δz and the increase of the time step Δt, the speed-up ratio of ICMP(3)-GS relative to GS has a tendency to increase. In Fig. 4.10, the transient numerical solutions obtained by the ICMP(m)-GS method are also very consistent with the analytical solutions. This result further verifies that the proposed method has a faster convergence rate and a higher acceleration effect than the conventional methods.

4.3.4.3 2D Transient Unsaturated Infiltration

The geometry and boundary conditions for this example are illustrated in Fig. 2.2, where $L = 1$ m and $W = 1$ m. The soil was assumed to be silt. The parameters of soil were described by Liu et al. (2015) and included $\theta_s = 0.35$, $\theta_r = 0.14$, $\alpha = 8 \times 10^{-3}$, and $k_s = 9 \times 10^{-4}$ m/h. The boundary conditions are as follows:

Table 4.4 Running time for solving steady-state infiltration

Conditions		Running time (s)					
Δz (m)	Δt (h)	GS	SOR	ICMP(1)-GS	ICMP(2)-GS	ICMP(3)-GS	ICMP(4)-GS
0.05	0.2	6.25	1.23	0.54	0.28	0.18	0.15
	0.1	6.66	1.60	0.74	0.43	0.29	0.23
	0.05	7.13	1.96	1.12	0.67	0.52	0.39
0.025	0.2	72.2	6.08	4.29	1.49	0.86	0.59
	0.1	77.8	7.82	4.96	3.12	1.19	1.03
	0.05	82.6	13.1	7.18	3.08	2.16	1.40

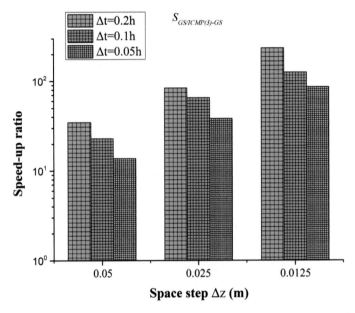

Fig. 4.9 Comparison of the speed-up ratios of ICMP(3)-GS relative to GS under different grid sizes

$$h(0, z, t) = h_d \qquad (4.37)$$

$$h(W, z, t) = h_d \qquad (4.38)$$

$$h(x, 0, t) = h_d \qquad (4.39)$$

$$h(x, L, t) = \ln\left(\left(1 - e^{\alpha h_d}\right)\sin\left(\frac{\pi x}{W}\right) + e^{\alpha h_d}\right)/\alpha \qquad (4.40)$$

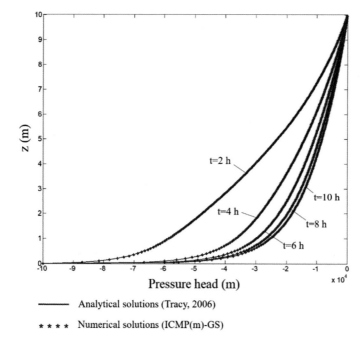

Fig. 4.10 Comparison of the numerical solutions obtained using ICMP(m)-GS with analytical solutions

The total simulation time and time step are set to 5 h and 0.1 h, respectively. The pressure profile computed using ICMP(m)-GS at $t = 5$ h is presented in Fig. 4.11, which is consistent with the analytical solution of this issue.

In Fig. 4.12, the condition number of the coefficient matrix can be greatly reduced using the proposed algorithm MP(m) and decreases with the increase of m. Figure 4.13a demonstrates the number of iterations of ICMP(m)-GS and GS. The number of iterations of ICMP(m)-GS is much smaller than that of the GS and decreases as m decreases. For $m > 2$, the number of iterations decreases slowly. Figure 4.13b indicates that the running time of ICMP(m)-GS is less than that of the GS, and the difference between the running time of GS and ICMP(3)-GS is about 10 times. Additionally, it can be found that when Δx, $\Delta z = 0.025$ m, the running time of ICMP(3)-GS is slightly shorter than that of ICMP(4)-GS. This is because when $m = 3$, the condition number of the coefficient matrix has been greatly reduced (Fig. 4.12). Thus increasing m cannot improve the computation efficiency. This result further demonstrates that the proposed method ICMP(m)-GS can achieve a significant acceleration compared to the GS, particularly at a smaller grid size.

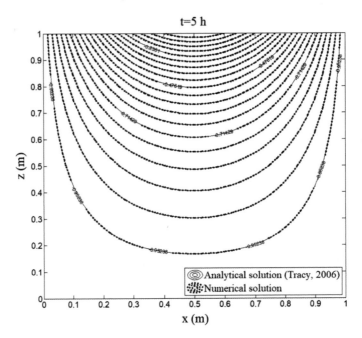

Fig. 4.11 Numerical solutions obtained using ICMP(m)-GS for 2D transient infiltration problem at 5 h

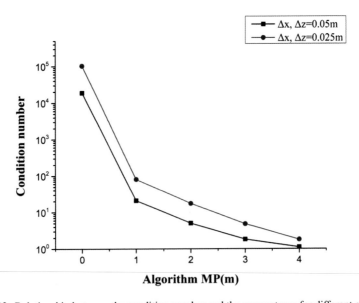

Fig. 4.12 Relationship between the condition number and the parameter m for different grid sizes

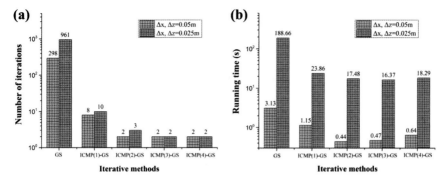

Fig. 4.13 Numerical results for this test: **a** number of iterations and **b** running time

4.4 Nonlinear Iterative Methods and Improvements

4.4.1 Newton and Picard Methods

In Fig. 4.14, the Richards' equation needs to be modified to satisfy rainfall infiltration into unsaturated soil slopes (Iverson 2000):

$$\frac{\partial}{\partial z}\left[K(h)\left(\frac{\partial h}{\partial z} + \cos\beta\right)\right] = \frac{\partial\theta}{\partial t} \tag{4.41}$$

where β is the slope angle.

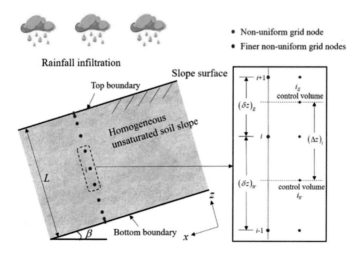

Fig. 4.14 Schematic map of control volume method based on non-uniform grid nodes and 1D infiltration model for homogeneous soil slope

Finite volume method is used to discretize Eq. (4.41). First, the interval on the z-axis is divided into N equal parts, and the simulation time is divided into M equal parts. Furthermore, Eq. (4.41) is integrated as follows (Patankar 1980):

$$\int_{i}^{i+\Delta z}\int_{t}^{t+\Delta t} \frac{\partial \theta}{\partial t} dt\, dz = \int_{t}^{t+\Delta t}\int_{i}^{i+\Delta z} \frac{\partial}{\partial z}\left[K(h)\left(\frac{\partial h}{\partial z} + \cos \beta \right) \right] dz\, dt \tag{4.42}$$

By simplification, one can gain:

$$\frac{K_{i+1/2}^{j}\left(h_{i+1}^{j} - h_{i}^{j} \right)}{\Delta z} - \frac{K_{i-1/2}^{j}\left(h_{i}^{j} - h_{i-1}^{j} \right)}{\Delta z} + \left(K_{i+1/2}^{j} - K_{i-1/2}^{j} \right)\cos \beta$$

$$= \Delta z C_{i}^{j-1/2}\frac{h_{i}^{j} - h_{i}^{j-1}}{\Delta t}, \quad 1 \le i \le N-1, \ 1 \le j \le M+1 \tag{4.43}$$

where i represents discrete nodes along the z-axis (except boundary nodes) and j represents time nodes. C is the specific moisture capacity, defined as $C(h) = \partial\theta/\partial h$; $K_{i+1/2}$ and $K_{i-1/2}$ represent the harmonic average of the hydraulic conductivity corresponding to adjacent nodes.

Newton and Picard iterative methods to solve Eq. (4.43) are a popular linearization technique. Newton method based on the one-dimensional Richards' equation in the form of pressure head (h) is expressed as:

$$\left(\mathbf{A}^{j,k} + \frac{\partial \mathbf{A}^{j,k}}{\partial \mathbf{h}^{j,k}}\mathbf{h}^{j,k} + \frac{\mathbf{C}^{j,k}}{\Delta t} + \frac{1}{\Delta t}\frac{\partial \mathbf{C}^{j,k}}{\partial \mathbf{h}^{j,k}}\left(\mathbf{h}^{j,k} - \mathbf{h}^{j-1} \right) - \frac{\partial \mathbf{F}^{j,k}}{\partial \mathbf{h}^{j,k}} \right)$$

$$\left(\mathbf{h}^{j,k+1} - \mathbf{h}^{j,k} \right) = \mathbf{F}^{j,k} - \mathbf{A}^{j,k}\mathbf{h}^{j,k} - \frac{\mathbf{C}^{j,k}}{\Delta t}\left(\mathbf{h}^{j,k} - \mathbf{h}^{j-1} \right) \tag{4.44}$$

where k is the number of iterations and $\mathbf{A}$ is a tridiagonal matrix, at node i, which is given by:

$$\mathbf{A}_{i} = \left[\frac{K_{i-1/2}^{j,k}}{\Delta z}, \ -\frac{K_{i-1/2}^{j,k}}{\Delta z} - \frac{K_{i+1/2}^{j,k}}{\Delta z}, \ \frac{K_{i+1/2}^{j,k}}{\Delta z} \right] \tag{4.45}$$

$\mathbf{C}$ and $\mathbf{F}$ can be written as follows at node i:

$$C_{i} = -\Delta z C_{i}^{j-1/2,k} \tag{4.46}$$

$$\mathbf{F}_{i} = \left(K_{i-1/2}^{j,k} - K_{i+1/2}^{j,k} \right)\cos \beta \tag{4.47}$$

The modified Picard method (PI), also known as the fixed-point method (Schrefler and Zhan 1993), can be expressed as:

$$\left(\mathbf{A}^{j,k} + \frac{\mathbf{C}^{j,k}}{\Delta t}\right)\left(\mathbf{h}^{j,k+1} - \mathbf{h}^{j,k}\right) = \mathbf{F}^{j,k} - \mathbf{A}^{j,k}\mathbf{h}^{j,k} - \frac{\mathbf{C}^{j,k}}{\Delta t}\left(\mathbf{h}^{j,k} - \mathbf{h}^{j-1}\right) \quad (4.48)$$

It can be seen from Eq. (4.48) that the modified Picard method is a simplification of the Newton method, ignoring the derivative term in the Jacobian matrix. In addition, Eq. (4.48) can be further simplified into the standard Picard format:

$$\left(\mathbf{A}^{j,k} + \frac{\mathbf{C}^{j,k}}{\Delta t}\right)\mathbf{h}^{j,k+1} = \mathbf{F}^{j,k} + \frac{\mathbf{C}^{j,k}}{\Delta t}\mathbf{h}^{j-1} \quad (4.49)$$

Newton's method is quadratic convergent (Li 1993), but needs to calculate the first derivatives of matrices $\mathbf{A}$ and $\mathbf{C}$ and vector $\mathbf{F}$, and the modified Picard method only converges linearly (Li 1993). Due to Newton's method increases the algebraic complexity and computational cost of constructing derivative terms in Jacobian matrices, thus modified Picard method has been widely investigated for h-based and mixed form of Richards' equations. Furthermore, there is still a large computational cost due to the need to recalculate the iteration matrix for each iteration in the Picard method (Farthing and Ogden 2017). For some unfavorable numerical conditions, such as infiltration into dry soils and/or layered soils with very different permeability coefficients, this results in numerical difficulties or even non-convergence of conventional Picard methods (Zha et al. 2017).

Normally, the uniform discrete process does not have good numerical convergence under some unfavorable numerical conditions, such as rainfall infiltration into dry soils (Zha et al. 2017). Therefore, a non-uniform grid in the form of Chebyshev is adopted (Wu et al. 2020). The coordinates in the two-layer soils in Fig. 4.14 can be expressed as follows:

$$z_{1i} = \cos(i\pi/N_1) \times \frac{L_1}{2} + \frac{L_1}{2}, \quad i = N_1, N_1 - 1, \ldots 0 \quad (4.50)$$

$$z_{2i} = \cos(i\pi/N_2) \times \frac{L_2}{2} + \frac{L_2}{2} + L_1, \quad i = N_2 - 1, N_2 - 2, \ldots 0 \quad (4.51)$$

where L_1 and L_2 are the height of soil layers 1 and 2, respectively; i denotes non-uniform grid nodes; and N_1 and N_2 represent the number of nodes in layers 1 and 2, respectively. In addition, for the convenience of comparison and analysis, one can have $N = N_1 + N_2$. The discretized equation is now derived by integrating Eq. (4.41) over the non-uniform finite volume in Fig. 4.14 and over the time interval from t to $t + \Delta t$. Thus,

$$\int_{i_W}^{i_E} \int_{t}^{t+\Delta t} \frac{\partial \theta}{\partial t} dt \, dz = \int_{t}^{t+\Delta t} \int_{i_W}^{i_E} \frac{\partial}{\partial z}\left[K(h)\left(\frac{\partial h}{\partial z} + \cos\beta\right)\right] dz \, dt \quad (4.52)$$

Furthermore, this chapter can obtain

$$\frac{K_{i_E}^{j,k}\left(h_{i+1}^{j,k+1}-h_i^{j,k+1}\right)}{(\delta z)_E} - \frac{K_{i_W}^{j,k}\left(h_i^{j,k+1}-h_{i-1}^{j,k+1}\right)}{(\delta z)_W} + \left(K_{i_E}^{j,k}-K_{i_W}^{j,k}\right)\cos\beta$$

$$= C_i^{j-1/2,k}(\Delta z)_i \frac{h_i^{j,k+1}-h_i^{j-1}}{\Delta t} \tag{4.53}$$

According to Eqs. (4.43) and (4.53), the matrix form of the standard Picard iterative method can be abbreviated as follows:

$$\mathbf{A}\left(\mathbf{h}^{j,k}\right)\mathbf{h}^{j,k+1} = \mathbf{f}\left(\mathbf{h}^{j,k},\mathbf{h}^{j-1}\right) \tag{4.54}$$

For the solution of Eq. (4.54), the system of linear equations is first derived and then solved by the basic linear solution methods. The Gaussian elimination method is usually effective. After solving Eq. (4.54) for the first time, the coefficient matrix in Eq. (4.54) is recalculated using this first solution, and the new linear equation is solved. The iteration process is terminated when the convergence measure is satisfied using l_∞ norm as:

$$\frac{\left\|\mathbf{h}^{j,k+1}-\mathbf{h}^{j,k}\right\|_\infty}{\left\|\mathbf{h}^{j,k}\right\|_\infty} < \varepsilon \tag{4.55}$$

The standard Picard method according to Eq. (4.54) can be abbreviated as N-PI. Generally, the Picard method combined with non-uniform grids has a slower convergence rate and lower calculation efficiency. To improve the convergence rate of N-PI, an improved Picard method based on non-uniform two-grid correction scheme (NTG-PI) is proposed in this book. In addition, the improved Picard method by the adaptive relaxation method (NAR-PI) is adopted for the comparative study.

4.4.2 Improved Picard Method

4.4.2.1 Adaptive Relaxed Picard Iterative Method (NAR-PI)

The adaptive relaxation method can effectively improve the computational efficiency of the Picard method (Smedt et al. 2010). First, when $k > 1$, the adaptive relaxation method can be expressed in the following form:

$$\mathbf{h}^{k+1} \leftarrow \mathbf{h}^k + \lambda_k\left(\mathbf{h}^{k+1}-\mathbf{h}^k\right); \quad \lambda_k \in (0,2) \tag{4.56}$$

where λ_k is the relaxation coefficient of the kth iteration, and its size can be adjusted according to the generalized angle between the current iteration step increment $\Delta\mathbf{h}$ and the previous iteration step increment $\Delta\overline{\mathbf{h}}$:

$$\delta = \arccos\left(\frac{\Delta\mathbf{h} \cdot \Delta\overline{\mathbf{h}}}{\|\Delta\mathbf{h}\|\|\Delta\overline{\mathbf{h}}\|}\right) \tag{4.57}$$

The angle δ in Eq. (4.57) represents the convergence trend of the numerical solution. An acute angle indicates better convergence, and an obtuse angle indicates oscillation. Therefore, when δ is an acute angle, λ_k should increase; when δ is an obtuse angle, λ_k should be appropriately reduced. After preliminary trial calculation, better calculation results can be obtained in the following forms (Smedt et al. 2010):

$$\lambda_k = \begin{cases} \sqrt{2}\lambda_{k-1}, & \delta < \pi/4 \\ \lambda_{k-1}, & \pi/4 \leq \delta < \pi/2 \\ \lambda_{k-1}/\sqrt{2}, & \delta \geq \pi/2 \end{cases} \tag{4.58}$$

Therefore, the adaptive relaxed Picard iterative method based on the non-uniform grid (NAR-PI) can be summarized as follows:

(1) Given the initial solution $\mathbf{h}_0$, and let $k = 1$.
(2) Solve Eq. (4.54) and calculate the current iteration increment $\Delta\mathbf{h}$.
(3) Calculate the relaxation coefficient λ_k of the kth iteration: if $k = 1$, then $\lambda_k = 1$; otherwise, read the previous iteration increment $\Delta\overline{\mathbf{h}}$ and relaxation coefficient λ_{k-1}, and obtain λ_k according to Eqs. (4.57)–(4.58).
(4) Equation (4.56) is used to modify the calculated result under the current iteration step.
(5) If Eq. (4.55) is satisfied, the iteration stops; otherwise, go to step 2.

4.4.2.2 Improved Picard Method Based on the Two-Grid Correction Scheme (NTG-PI)

The algebraic multigrid method can improve the computational efficiency of nonlinear iterative methods (Wang and Schrefler 2003). In this chapter, the non-uniform two-grid correction scheme (NTG) is adopted to improve the computational efficiency of the classical Picard method. In Fig. 4.15, the non-uniform two-grid correction scheme is described.

First, the solution vectors of the fine and coarse grids are assumed to be $\mathbf{h}^f$ and $\mathbf{h}^c$, respectively. In NTG, the Gauss–Seidel iterative method is usually used as the basic linear solution method instead of the Gaussian elimination method. The implementation steps of NTG-PI are expressed in detail as follows (Zhu et al. 2022a, b):

(1) Relax μ times using Gauss–Seidel relaxation for the fine grid with initial vector $\mathbf{h}^{f0}$.
(2) Compute the fine-grid residual $\mathbf{r}^f = \mathbf{f}(\mathbf{h}^{f\mu}) - \mathbf{A}(\mathbf{h}^{f\mu})\mathbf{h}^{f\mu}$ and restrict it to the coarse grid by $\mathbf{r}^c = \mathbf{R}\mathbf{r}^f$ and $\mathbf{h}^c = \mathbf{R}\mathbf{h}^{f\mu}$, where $\mathbf{R}$ is the restriction operator from fine to coarse grids.

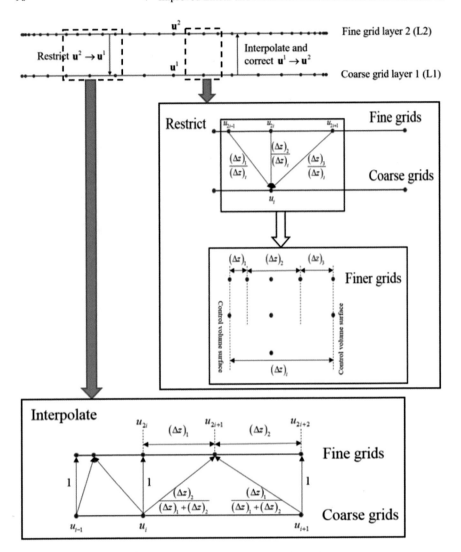

Fig. 4.15 Schematic drawing of non-uniform two-grid correction scheme

(3) Solve the following residual equation for the coarse grid:

$$\mathbf{A}(\mathbf{h}^c)\mathbf{e}^c = \mathbf{r}^c \tag{4.59}$$

(4) Interpolate the coarse-grid error $\mathbf{e}^c$ to the fine grid and correct the fine-grid approximation solution by:

$$\mathbf{h}^{f\mu} \leftarrow \mathbf{h}^{f\mu} + \mathbf{I} \cdot \mathbf{e}^c \tag{4.60}$$

where $\mathbf{I}$ denotes the interpolation operator from coarse to fine grids.

(5) Relax μ times using Gauss–Seidel relaxation for the fine grid with corrected vector $\mathbf{h}^{f\mu}$, where μ usually takes a small natural number. μ is set to 1 in this chapter.

If $\mathbf{h}^{f\mu}$ converges according to Eq. (4.55), then stop; otherwise, go to step 1. Steps 1 and 5 can eliminate the high-frequency components of the iteration error, while steps 2–4 can eliminate the remaining low-frequency smooth components. Through multiple cycles of the above steps, it is expected to eliminate the iteration error as soon as possible. And this makes the iterative method more efficient. In addition, it is needed to further define the interpolation ($\mathbf{I}$) and restriction ($\mathbf{R}$) operators. In Fig. 4.15, the restriction operator for nodes i can be expressed as:

$$u_i = \frac{(\Delta z)_1}{(\Delta z)_i} u_{2i-1} + \frac{(\Delta z)_2}{(\Delta z)_i} u_{2i} + \frac{(\Delta z)_3}{(\Delta z)_i} u_{2i+1} \tag{4.61}$$

The interpolation operator can be written as:

$$u_{2i} = u_i \tag{4.62}$$

$$u_{2i+1} = \frac{(\Delta z)_2}{(\Delta z)_1 + (\Delta z)_2} u_i + \frac{(\Delta z)_1}{(\Delta z)_1 + (\Delta z)_2} u_{i+1} \tag{4.63}$$

$$u_{2i+2} = u_{i+1} \tag{4.64}$$

4.4.3 Numerical Tests—1D Transient Unsaturated Infiltration

This test simulates the 1D transient infiltration in homogeneous soils to verify the effectiveness of the proposed schemes. The mathematical model is described in Fig. 4.14. The boundary conditions and the parameters are kept unchanged. The total simulation time is 10 h. The parameter μ is set to 1. Additionally, the number of nodes is taken as 100, 200, and 400, and the time steps are taken as 0.1 h, 0.05 h, and 0.01 h, respectively.

Figure 4.16a represents the maximum relative error (MRE) obtained by different methods under different time steps when the number of nodes $N = 200$. It can be seen that the range of MRE obtained by NTG-PI is between 0.35 and 4.6%. When t is less than 6 h, the MRE decreases as the time step decreases. However, the range of MRE obtained by PI is between 2.7 and 72%, which increases with the increase of t and is much larger than that obtained by NTG-PI. Figure 4.16b shows the MRE obtained by different methods under different number of nodes when the time step $\Delta t = 0.01$ h. It can be found that the MRE obtained by PI increases over time t. When

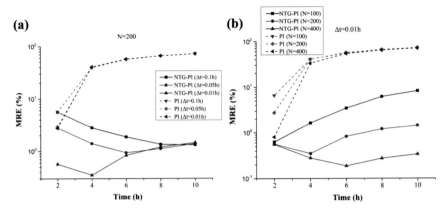

Fig. 4.16 Comparison of the maximum relative error of different methods under different numerical conditions: **a** time step and **b** number of nodes N

t is greater than 2 h, the error becomes larger and larger, while the MRE obtained by NTG-PI only ranges from 0.18 to 8%. In addition, the MRE of the two methods decreases as the number of nodes N increases.

In Fig. 4.17, the numerical solutions obtained by the two methods under $\Delta t = 0.01$ h and $N = 200$ are compared with the exact solutions. It can be found that there is a great deviation between the numerical solution obtained by PI and the exact solution, particularly after the time is greater than 4 h (Fig. 4.17a). On the contrary, the numerical solution obtained by NTG-PI is very consistent with the exact solution, and there is no large relative error as the time increases (Fig. 4.17b). When $t = 10$ h, the RSE of NTG-PI decreases with the increase of N and the decrease of Δt in Table 4.5, which seems to be better than that of PI. And the RE of NTG-PI is much smaller than that of PI, and the difference is about 100 times. In addition, Fig. 4.18 depicts the convergence rate of the proposed methods. It can be found that the convergence rate of the proposed method NTG-PI is faster than that of the conventional method N-PI. As the number of nodes N changes, the convergence rate of NTG-PI does not change much.

In Table 4.6, the speed-up ratio of the improved method NTG-PI relative to N-PI is much greater than that of NAR-PI relative to N-PI, with a difference of nearly 10 times. Since the parameter μ is only 1, the computational efficiency of the proposed NTG-PI is higher. This test indicates that the proposed NTG-PI can obtain higher numerical accuracy with a smaller number of nodes and has a faster convergence rate while reducing the computational cost.

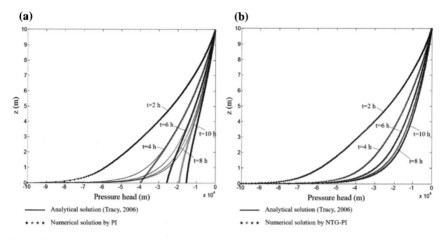

Fig. 4.17 Comparison of the numerical solutions obtained using different methods with analytical solutions: **a** PI and **b** NTG-PI

Table 4.5 Numerical accuracy of transient infiltration in homogeneous soils at $t = 10$ h

Conditions		RSE		RE (%)	
N	Δt (h)	PI	NTG-PI	PI	NTG-PI
100	0.1	0.25	1.2e−02	18.33	0.42
	0.05	0.25	1.1e−02	18.84	0.035
	0.01	0.26	1.1e−02	19.24	0.27
200	0.1	0.23	7.9e−03	16.17	0.69
	0.05	0.23	4.2e−03	16.69	0.30
	0.01	0.24	2.1e−03	17.10	0.0033
400	0.1	0.19	7.9e−03	12.24	0.74
	0.05	0.19	3.9e−03	12.76	0.36
	0.01	0.19	8.9e−04	13.18	0.056

4.5 Conclusions

This chapter first discusses the classical linear stationary iterative methods, such as Jacobi iterative method, Gauss–Seidel iterative method, and SOR iterative method. Then, combined with the preprocessing technology and error correction method, a series of improved iterative methods is proposed. The proposed methods have better computational performance in the simulation of rainfall infiltration and can obtain good application results. The main conclusions are as follows:

(1) Chebyshev semi-iterative method with polynomial preconditioner (P-CSIM) is developed. Compared with conventional iterative methods such as Jacobi and Gauss–Seidel methods, P-CSIM has less iterations and calculation time in

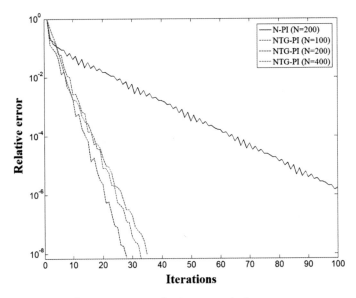

Fig. 4.18 Comparison of convergence rates for the proposed schemes

Table 4.6 Speed-up ratio of two improved Picard methods relative to N-PI

Time step Δt (h)	Speed-up ratio	
	$S_{\text{N-PI/NAR-PI}}$	$S_{\text{N-PI/NTG-PI}}$
0.1	3.33	46.69
0.05	1.72	24.99
0.01	1.07	14.12

simulating rainfall. Numerical results indicate significant speed-up, at least 50% increase compared to GSIM, and higher computational efficiency compared to CSIM and JIM.

(2) In the numerical solution of unsaturated infiltration, the numerical discrete condition plays a significant role in the convergence rate of the numerical solution, particularly under the small space step and large time step. Conventional iterative methods such as Jacobi and GS tend to converge very slowly and have low computational efficiency. For the accelerated improvement methods, the integral correction method uses the solution vector error correction to make the iterative method obtain a faster convergence rate, while the multistep preconditioning method preprocesses the original system of linear equations into a more easily solved system of the linear equations, making the subsequent iterative method more effective.

(3) The proposed ICMP(m)-GS is a mixture of the integral correction method and the multistep preconditioning method. The adjustment of the parameters c and m can not only greatly improve the ill-condition of the linear equations derived

from the linearized Richards' equation, but also obtain higher numerical accuracy. Numerical results show that compared with the conventional methods GS and SOR and the improved methods IC(c)-GS and MP(m)-GS, ICMP(m)-GS can improve the convergence rate to a greater degree and has higher computational efficiency and calculation accuracy. This method is relatively simple, and it has a good application prospect for simulating unsaturated infiltration.

(4) In the conventional Picard method, the Gaussian elimination method is usually used to solve the system of linear equations. Although the solution obtained by this method is theoretically accurate, with the increase of the number of discrete nodes N, the calculation amount increases in the order of N^3, and the computational efficiency significantly reduced. In the improved method NTG-PI, the Gauss–Seidel iterative method is used instead of the Gaussian elimination method as the basic method, and its solution speed is faster. Numerical results demonstrate that the numerical solution obtained by the NTG-PI is in good agreement with the analytical solution under extremely dry initial conditions. Compared with the conventional Picard method, NTG-PI can achieve higher numerical accuracy with fewer discrete nodes, while simulating rainfall infiltration with faster convergence.

References

Arioli M, Scott J (2014) Chebyshev acceleration of iterative refinement. Numer Algorithms 66(3):591–608

Bai ZZ, Golub GH, Pan JY (2004) Preconditioned Hermitian and skew-Hermitian splitting methods for non-Hermitian positive semidefinite linear systems. Numer Math 98(1):1–32

Bai ZZ, Benzi M, Chen F (2010) Modified HSS iteration methods for a class of complex symmetric linear systems. Computing 87(3–4):93–111

Benzi M (2002) Preconditioning techniques for large linear systems: a survey. J Comput Phys 182(2):418–477

Both JW, Kumar K, Nordbotten JM, Radu FA (2018) Anderson accelerated fixed-stress splitting schemes for consolidation of unsaturated porous media. Comput Math Appl 77:1479–1502

Brenner K, Cancès C (2017) Improving Newton's method performance by parametrization: the case of the Richards equation. SIAM J Numer Anal 55(4):1760–1784

Briggs WL, Henson VE, McCormick SF (2000) A multigrid tutorial, 2nd edn. SIAM, Philadelphia, PA

Casulli V, Zanolli P (2010) A nested Newton-type algorithm for finite volume methods solving Richards' equation in mixed form. SIAM J Sci Comput 32(4):2255–2273

Celia M, Bouloutas E, Zarba R (1990) A general mass-conservative numerical solution for the unsaturated flow equation. Water Resour Res 26:1483–1496

Dehghan M, Mohammadi-Arani R (2016) Generalized product-type methods based on bi-conjugate gradient (GPBiCG) for solving shifted linear systems. Comput Appl Math 36(4):1591–1606

Dehghan M, Shirilord A (2019) Accelerated double-step scale splitting iteration method for solving a class of complex symmetric linear systems. Numer Algorithms 83(1):281–304

Deng B, Wang J (2017) Saturated-unsaturated groundwater modeling using 3D Richards equation with a coordinate transform of nonorthogonal grids. Appl Math Model 50(10):39–52

Farthing MW, Ogden FL (2017) Numerical solution of Richards' equation: a review of advances and challenges. Soil Sci Soc Am J 81(6):1257–1269

Hagemam LA, Young DM (1981) Applied iterative methods. Academic Press, New York

Illiano D, Pop IS, Radu FA (2021) Iterative schemes for surfactant transport in porous media. Comput Geosci 25(2):805–822

Iverson RM (2000) Landslide triggering by rain infiltration. Water Resour Res 36(7):1897–1910

Lehmann F, Ackerer P (1998) Comparison of iterative methods for improved solutions of the fluid flow equation in partially saturated porous media. Transport Porous Med 31(3): 275–292.

Li CW (1993) A simplified Newton iteration method with linear finite elements for transient unsaturated flow. Water Resour Res 29(4):965–971

Liu CY, Ku CY, Huang CC (2015) Numerical solutions for groundwater flow in unsaturated layered soil with extreme physical property contrasts. Int J Nonlinear Sci Numer Simul 16(7):325–334

List F , Radu FA (2016) A study on iterative methods for solving richards' equation. Computat Geosci 20(2):341-235

Lott PA, Walker HF, Woodward CS, et al (2012) An accelerated Picard method for nonlinear systems related to variably saturated flow. Adv Water Resour 38: 92–101

Mitra K, Pop IS (2019) A modified L-scheme to solve nonlinear diffusion problems. Comput Math Appl 77:1722–1738

Patankar SV (1980) Numerical heat transfer and fluid flow. CRC Press

Paniconi C, Putti M (1994) A comparison of Picard and newton iteration in the numerical solution of multidimensional variably saturated flow problems. Water Resour Res 30: 3357-3374

Radu FA, Pop IS, Knabner P (2006) On the convergence of the Newton method for the mixed finite element discretization of a class of degenerate parabolic equation. In: Numerical mathematics and advanced applications. Springer, pp 1194–1200

Schrefler BA, Zhan X (1993) A fully coupled model for water flow and airflow in deformable porous media. Water Resour Res 29(1):155–167

Smedt BD, Pattyn F, Groen PD (2010) Using the unstable manifold correction in a Picard iteration to solve the velocity field in higher-order ice-flow models. J Glaciol 56(196):257–261

Su X, Zhang M, Zou D et al (2022) Numerical scheme for solving the Richard's equation based on finite volume model with unstructured mesh and implicit dual-time stepping. Comput Geotech 147:104768

Walker HF, Ni P (2011) Anderson acceleration for fixed-point iterations. SIAM J Numer Anal 49(4):1715–1734

Wang XC, Schrefler BA (2003) Fully coupled thermo-hydro-mechanical analysis by an algebraic multigrid method. Eng Comput 20(2):211–229

Wang K, Zhang J (2003) MSP: a class of parallel multistep successive sparse approximate inverse preconditioning strategies. SIAM J Sci Comput 24(4):1141–1156

Wu LZ, Zhu SR, Peng JB (2020a) Application of the Chebyshev spectral method to the simulation of groundwater flow and rainfall-induced landslides. Appl Math Model 80:408–424

Wu LZ, Huang RQ, Li X (2020b) Hydro-mechanical analysis of rainfall-induced landslides. Springer

Zeng JC, Zha YY, Yang JZ (2018) Switching the Richards' equation for modeling soil water movement under unfavorable conditions. J Hydrol 563:942–949

Zha Y, Yang J, Yin L et al (2017) A modified Picard iteration scheme for overcoming numerical difficulties of simulating infiltration into dry soil. J Hydrol 551:56–69

Zhu SR, Wu LZ, Peng JB (2020) An improved Chebyshev semi-iterative method for simulating rainfall infiltration in unsaturated soils and its application to shallow landslides. J Hydrol 590:125157

Zhu SR, Wu LZ, Huang J (2022a) Application of an improved P(m)-SOR iteration method for flow in partially saturated soils. Comput Geosci 26:131–144

Zhu SR, Wu LZ, Song XL (2022b) An improved matrix splititeration method for analyzing underground water flow. Eng Comput. https://doi.org/10.1007/s00366-021-01551-z

Zhu S R, Wu L Z, Cheng P, Zhou JT (2022c) Application of modified iterative method to simulate rainfall infiltration in unsaturated soils. Comput Geotech 148: 104816

Chapter 5
Slope Stability Analysis Based on Analytical and Numerical Solutions

5.1 Introduction

Infiltration into soil slopes is a fundamental concern in civil engineering. Rainfall infiltration leads to changes in pore-water pressure and reduces matric suction in soils, making it one of the main triggers of slope failure (Rahimi et al. 2010; Ali et al. 2014; Wu et al. 2020). Slope instabilities caused by water infiltration are called rainfall-induced landslides (Xu and Zhang 2010; Wu et al. 2020).

There are three methods that can be taken to investigate the effect of rainfall infiltration on pore-water pressure or pressure head profile and hence on unsaturated soil slope stabilities. The approaches include numerical simulation, field monitoring, and analytical analysis (Zhan et al. 2013). A number of numerical studies have been conducted to investigate the hydrodynamic behaviors of soil slopes due to rainfall infiltration and to estimate the influence on the slope stability (Ng and Shi 1998; Iverson 2000; Collins and Znidarcic, 2004; Kim et al. 2004; Zhang et al. 2005; Garcia et al. 2011; Ali et al. 2014). Factors affecting the soil slope stability due to rainwater infiltration comprise the rainfall characteristics, the saturated permeability coefficient, the geometry of the slope, and the boundary and initial soil moisture conditions (Ali et al. 2014). Laboratory and field experiments have been performed to investigate variations in matric suction due to rainfall infiltration to improve the understanding of the mechanism of rainfall-induced soil slope failures (Wu et al. 2015, 2017, 2020). Field monitoring is helpful in the study of the effects of rainfall infiltration on the slope stability, but it is a costly procedure (Zhan et al. 2013).

The analytical methods involve theoretical infiltration and approximate infiltration models (Wu et al. 2022). In a theoretical infiltration model, the partial differential equation describing water infiltration in soils (i.e., the Richards' equation) is proposed based on continuum mechanics, and the solution can be obtained through integral transformation or numerical methods (Srivastava and Yeh 1991; Zhu et al. 2020, 2022).

© The Author(s) 2023
L. Wu and J. Zhou, *Rainfall Infiltration in Unsaturated Soil Slope Failure*,
SpringerBriefs in Applied Sciences and Technology,
https://doi.org/10.1007/978-981-19-9737-2_5

The analytical method requires that certain assumptions be made regarding the closed-form equations that are derived. If the assumptions can be shown to be reasonable, then the analytical methodology becomes a simple and practical tool for studying possible pressure head distribution and the factor of safety of the slope.

Coupled poromechanical approaches incorporating the unsaturated soils have been employed to analytically and numerically examine the hydrodynamic behaviors of partially or completely saturated slopes due to rainfall (Chen et al. 2005). These coupled poromechanical studies are in view of various constitutive models, which represent the mechanical response of the unsaturated soils using nonlinear elastic stress–strain relationships (Cho and Lee 2001; Zhang et al. 2005) or elasto-viscoplastic and elastoplastic models. The coupled hydromechanical approach has been combined with a slope stability analysis according to finite element methods to examine the response of an unsaturated soil slope subjected to a rainstorm and to consider the variability in the soil properties (Zhang et al. 2005). Consequently, using the improved method proposed in the previous chapters, the infiltration equation of the unsaturated soil slope is solved, and the numerical and analytical solutions are applied to the slope model to study the effect of rainfall infiltration on the slope stability.

The practical infiltration process is very complicated and affected by many factors such as soil profile and rainfall conditions and becomes difficult to be described accurately with theoretical formulations (D'Aniello et al. 2019; Srivastava et al. 2020).

Due to the nonlinearity of differential equations, numerical software packages are frequently needed to solve the complex problem of infiltration for predicting moisture movement and pore-water pressure change in unsaturated soils (Masoudian et al. 2019; Yang et al. 2018; Zhang et al. 2016).

Numerical approaches often suffer from convergence and mass balance problems and also are inefficient and expensive in many conditions. To provide simplified solutions to the complex infiltration issue, a number of theoretical models based on the wetting front have been proposed (Iverson 2000; Conte and Troncone, 2008). A series of analytical approaches (Srivastava and Yeh 1991; Iverson 2000; Wu et al. 2022) was developed to calculate the pressure head change during rainfall. Unfortunately, these models can only be applied under the assumption that the rainfall intensity is a constant. However, in practice, there are often cases where the rainfall intensity is a variable function depending on duration.

The objective of this chapter is to study the application of improved numerical methods to the slope stability assessment and to discuss the influence of various factors on the slope stability. Combined with the monitoring data of slope pore-water pressure in the Tung Chung area of Hong Kong, the book carried out a comparative study to verify the accuracy and practicability of the proposed improved method. The Xiaoba landslide in Guizhou, China, a typical rainfall-induced landslide, was investigated using numerical methods.

5.2 Application of Unsaturated Soil Slope Stability Under Rainfall

5.2.1 Application of Chebyshev Spectral Method in Slope Stability Analysis

For unsaturated soil slopes (Fig. 5.1), the governing equations are modified as follows:

$$\frac{\partial}{\partial z}\left[K_z(h)\left(\frac{\partial h}{\partial z} + \cos\beta\right)\right] = \frac{\partial\theta}{\partial t} \tag{5.1}$$

where β is the slope angle.

According to modified Eq. (2.34), the governing equation considering the hydromechanical coupling in soil slopes can be obtained:

$$\frac{\partial}{\partial z^*}\left[K(h)\frac{\partial h}{\partial z^*}\right] + \frac{\partial K(h)}{\partial z^*}\cos\beta = C(h)\frac{\partial h}{\partial t} + \frac{\alpha_c\gamma_w(\theta - \theta_r)}{(\theta_s - \theta_r)F}\frac{\partial h}{\partial t} \tag{5.2}$$

The analytical solution of the infiltration equation along the z-axis can be expressed as:

$$h_t^*(z, t) = \frac{2(1 - e^{\alpha h_d})}{Lc}e^{\alpha\cos\beta(L-z)/2}\sum_{m=1}^{\infty}(-1)^m\left(\frac{\lambda_m}{\mu_m}\right)\sin(\lambda_m z)e^{-\mu_m t} \tag{5.3}$$

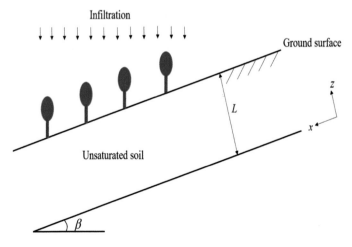

Fig. 5.1 Schematic diagram of a homogeneous unsaturated slope

$$h_s^*(z) = \left(1 - e^{\alpha h_d}\right)\frac{1 - e^{-\alpha \cos \beta z}}{1 - e^{-\alpha \cos \beta L}} \tag{5.4}$$

This application example simulated transient infiltration in an unsaturated slope using the Chebyshev spectral method (CSM). The slope thickness (L) and angle (β) are assumed to be 10 m and 33°, respectively (Fig. 5.1). The initial condition was unchanged at $h(z, t = 0) = -10$ m. The upper and lower boundary conditions of the soil can be expressed as $h(z = 0) = -10$ m and $h(z = L) = 0$.

For this example, this chapter selected sandy soil and silty loam. The experimental data (Brooks and Corey 1964; Lu and Likos 2004) and fitting curve of the two soils are shown in Fig. 5.2. The fitted parameters including k_s, θ_s, θ_r, and α are listed in Table 5.1. The total simulation time was 2 h and $N = 40$.

The pressure head calculated using the CSM was introduced into the infinite slope stability analysis, and the factor of safety (F_s) could be obtained (Iverson 2000; Liu et al. 2017):

$$F_s = \frac{c'}{z\gamma_t \cos \beta \sin \beta} + \frac{\tan \phi'}{\tan \beta} - \frac{h\gamma_w \tan \phi'}{z\gamma_t \cos \beta \sin \beta} \tag{5.5}$$

where c' denotes the effective cohesion, ϕ' denotes the effective friction angle, γ_t denotes the unit weight of soil, γ_w denotes the unit weight of water, and z denotes the soil thickness. It was assumed that the unit weight of soil was 19 kN/m³, and the effective cohesiveness and effective friction angle of sandy soil and silty loam were 0 kPa, 32° and 10 kPa, 20°, respectively (Lu and Likos 2004).

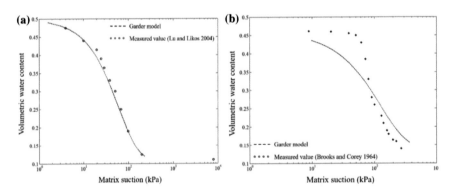

Fig. 5.2 Fitting curves of the SWCC: **a** sandy soil and **b** silty loam

Table 5.1 Parameters of unsaturated soils for analysis

Soil type	k_s (m/h)	α	θ_s	θ_r
Sandy soil	6.0×10^{-2}	1.6×10^{-2}	0.50	0.11
Silty loam	1.5×10^{-3}	8.0×10^{-3}	0.46	0.14

The revised F_s for analyzing slope stability can be obtained as (Liu et al. 2017):

$$F_s = \frac{c'}{z\gamma_t \cos \beta \sin \beta} + \frac{\tan \phi'}{\tan \beta} - \frac{h\gamma_w \left(\frac{\theta - \theta_r}{\theta_s - \theta_r}\right) \tan \phi'}{z\gamma_t \cos \beta \sin \beta} \tag{5.6}$$

The profiles of the pressure head of the two soils over time were compared with the analytical solutions. Figure 5.3 shows that the computed results were in good agreement with the analytical solutions. The numerical findings for sandy soil and silty loam indicate that the accuracy of the absolute error can reach the order of 10^{-5} and 10^{-6}, respectively (Fig. 5.4).

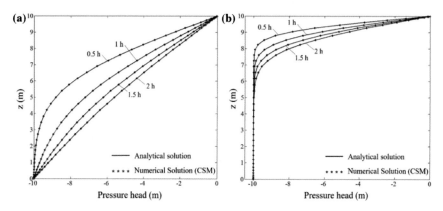

Fig. 5.3 Results comparison with the exact solution for unsaturated soil: **a** sandy soil and **b** silty loam

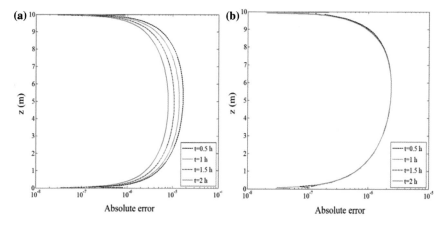

Fig. 5.4 Absolute error of the computed results with the analytical solutions for unsaturated soil: **a** sandy soil and **b** silty loam

Figure 5.5 demonstrates the computed results of F_s for different unsaturated soils. Additionally, the range of F_s for different soils is different, and F_s for silty loam is significantly smaller than that for sandy soil. Table 5.2 shows variations in F_s at different depths over the infiltration time for different soils. It can be seen that the factor of safety decreased with increasing depth and infiltration time. F_s for sandy soil was greater than 1.0 in Table 5.2, and F_s for silty loam was less than 1.0 at $z = 9.263$ m. Consequently, the stability of the slope of sandy soil was better than that of the slope of silty loam. The comparison of the computed results reveals that rainfall-induced landslides are closely related to soil types.

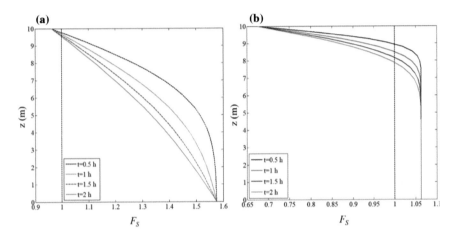

Fig. 5.5 Computed results of F_s for unsaturated soil: **a** sandy soil and **b** silty loam

Table 5.2 Variations in the factor of safety (F_s) at different depths over the infiltration time for different soils

Conditions		Factor of safety (F_s)			
Soil type	z (m)	0.5 h	1 h	1.5 h	2 h
Sandy soil	9.263	1.0828	1.0473	1.0315	1.0225
	8.247	1.2337	1.1590	1.1240	1.1038
	7.27	1.3519	1.2556	1.2071	1.1785
Silty loam	9.263	0.9361	0.8743	0.8423	0.8219
	8.247	1.0546	1.0233	0.9928	0.9677
	7.27	1.0627	1.0585	1.0480	1.0348

5.2.2 Application of P-CSIM in the Stability Analysis of Shallow Rainfall Slope

The proposed P-CSIM is used to solve the Richards' equation for the slope, then the numerical solution of the pressure head is substituted into the infinite slope model, and the stability of the slope is analyzed. The factor of safety of the slopes is calculated as Eqs. (5.5)–(5.6).

5.2.2.1 Homogeneous Soil Slope

The soil slope thickness is 150 cm, the slope angle is 33°, the initial condition for removing the boundary point is $h(z, t = 0) = -100$ cm, and the boundary conditions are $h(z = 0) = -100$ cm and $h(z = L) = 0$. The homogeneous unsaturated soil is assumed to be a sandy soil with a saturated permeability coefficient k_s, residual volumetric water content θ_r, and saturated volumetric water content θ_s of 9×10^{-2} cm/h, 0.08, and 0.43, respectively (Liu et al. 2017). The desaturation coefficient α is 1×10^{-2}. In addition, the unit weight, effective cohesiveness, and effective friction angle of the soil are 21.5 kN/m³, 4.6 kPa, and 30°, respectively (Lu and Likos 2004).

Figures 5.6 and 5.7 represent the computed profile of the factor of safety and the pressure head for the homogeneous soil slope. The suction (negative pressure head) in shallow parts of the slope decreases with the increase of rainfall time. In addition, it is found that F_s decreases with the increase of rainfall time at the shallow slope zone. The unstable slope is mostly 130–150 cm deep, which may be strength deterioration and softening of the shallow layer caused by rainfall infiltration (Zhuang et al. 2018).

5.2.2.2 Two-Layer Soil Slopes

For slopes of two-layer soils, the thickness of unsaturated Soil 1 (L_1) and Soil 2 (L_2) is assumed to be 130 cm and 20 cm, respectively (Fig. 5.8). The slope angle and boundary and initial conditions are the same as the slope for homogeneous soils. The physical parameter settings are listed in Table 5.3. It is assumed that Soil 1 is loess, and that the permeability coefficient of Soil 2 is greater than that of Soil 1.

The calculated results (Fig. 5.9) show that increase of the pressure head in layered unsaturated soils dominates slope stability. It can be found that the pressure head changes faster at the interface. Besides, because the permeability coefficient of the bottom layer is lower than that of the upper layer, the top layer (unsaturated Soil 2) may become unstable. Figure 5.10 demonstrates F_s profile of the two-layer unsaturated soil slope. It shows that unstable slope ranges from the interface to the soil surface after $t = 10$ h. Figure 5.11 indicates that F_s varies with duration at the interface ($z = 130$ cm) for unsaturated soil slopes. It is found that F_s is strongly affected by duration, and that the upper layer of slope becomes unstable after $t = 8.2$ h.

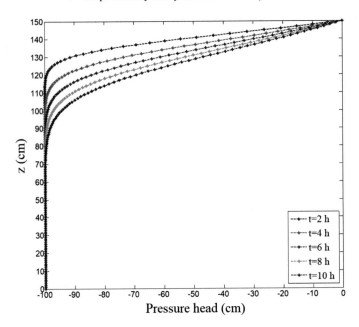

Fig. 5.6 Computed profile of pressure head for homogeneous unsaturated soil slopes

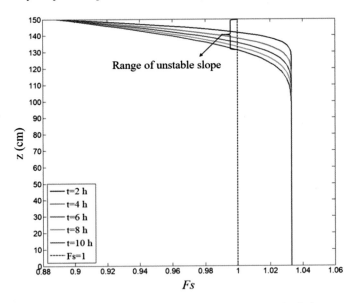

Fig. 5.7 Computed profile of safety factor for homogeneous unsaturated soil slopes

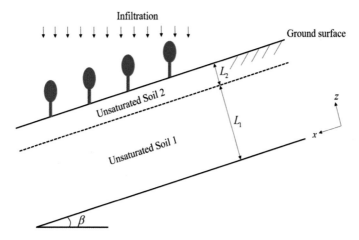

Fig. 5.8 Schematic diagram of two-layer soil slope model

Table 5.3 Parameters used in two-layer soil slopes

Parameters	k_s	θ_s	θ_r	α
Soil 1 (sand)	10^{-2}	0.43	0.08	1×10^{-2}
Soil 2 (silty loam)	10^{-1}	0.50	0.11	1×10^{-2}

Accordingly, the results illustrate that the pressure head caused by rainfall infiltration is closely related to the soil layer, and the interface of the soil layer plays a key role in the slope stability related to rainfall infiltration.

5.2.3 Application of Improved Picard Method into Unsaturated Slopes

In this example, this chapter investigates 1D transient infiltration into the two-layer soil slopes. The van Genuchten model is adopted here. In Fig. 5.8, the thickness of soil layer 1 (L_1) and soil layer 2 (L_2) is 3 m and 2 m, respectively, and the slope angle (β) is 33°. The governing equation is shown in Eq. (5.1). The VGM model parameters of the two soils are listed in Table 5.4 (Zha et al. 2013). The bottom boundary condition is groundwater table, defined as $h(z = 0, t) = 0$. The top boundary condition is given by (Wu et al. 2020):

$$\left[K \frac{\partial h}{\partial z} + K \right]_{z=L_1+L_2} = q_t, \quad t < t_1 \tag{5.7}$$

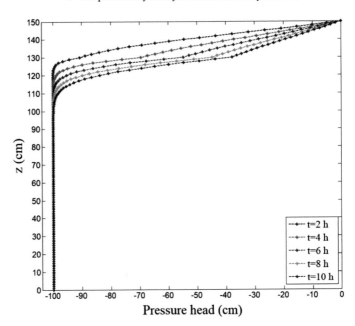

Fig. 5.9 Computed profile of pressure head for two-layer unsaturated soil slopes

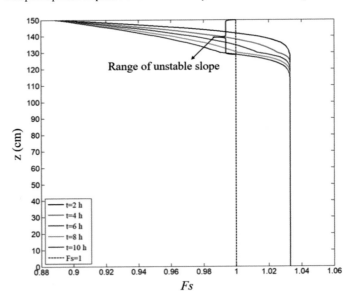

Fig. 5.10 Computed profile of safety factor for two-layer unsaturated soil slopes

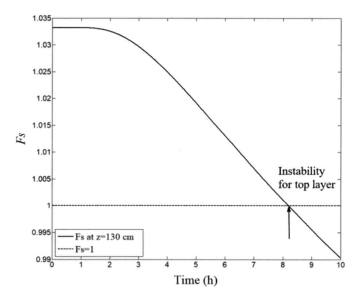

Fig. 5.11 Results of safety factor at the interface ($z = 130$ cm) for two-layer unsaturated soil slopes

Table 5.4 Parameter of van Genuchten model

Soil texture	k_s (m/h)	θ_s	θ_r	α	n
Glendale clay loam	5.46×10^{-3}	0.4686	0.106	1.04	1.3954
Berino loamy fine sand	2.25×10^{-1}	0.3658	0.0286	2.8	2.239

$$h_{|z=L_1+L_2} = h_t, \quad t \geq t_I \tag{5.8}$$

where t_I represents ponding time; q_t denotes the rainfall flux at the soil surface when the rainfall time is less than the ponding time; and h_t denotes constant pressure head after the ponding time. Here, q_t is assumed to be $K_{s2}/4$ and $h_t = 0$. The initial condition is given by $h(z, t = 0) = -z \times 10$ m.

The total simulation time and time step are taken to be 2 h and 0.01 h, respectively. Let the number of nodes be $N_1 = 60$ and $N_2 = 40$. Furthermore, NTG-PI is applied to solve Eq. (5.1), where $\mu = 1$ is adopted. The unit weight of the soil, the effective cohesion, and friction angle are 19.5 kN/m^3, 4.6 kPa, and 30°, respectively.

Figure 5.12 shows that the pressure head increases over duration, while the pressure head at the interface has a greater increase. Compared with NTG-PI, the pressure head obtained by traditional method PI has a greater increase at the interface as the rainfall time increases. This may underestimate the safety factor of the interface for two-layer soil slopes (Fig. 5.13). Moreover, the results illustrate that the soil layer structure will affect the distribution of the pressure head in the two-layer soil slopes during rainfall. When the hydraulic conductivity of the lower soil is lower than that of the upper one, it is easy to form ponding at the interface and induce landslides.

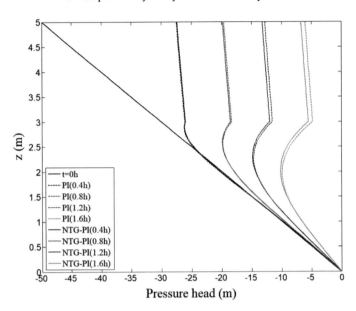

Fig. 5.12 Computed profile of pressure head for two-layer soil slopes

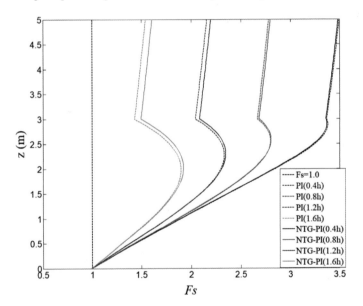

Fig. 5.13 Computed profile of F_s for two-layer soil slopes

Table 5.5 Parameters for different cases

Case	K_s (m/h)	α	β	q
1	3.6e−4	0.1	30	K_s
	3.6e−3	0.1	30	K_s
	3.6e−2	0.1	30	K_s
2	3.6e−3	0.01	30	$0.5\,K_s$
	3.6e−3	0.02	30	$0.5\,K_s$
	3.6e−3	0.05	30	$0.5\,K_s$
	3.6e−3	0.1	30	$0.5\,K_s$
3	3.6e−2	0.3	20	$0.5\,K_s$
	3.6e−2	0.3	30	$0.5\,K_s$
	3.6e−2	0.3	40	$0.5\,K_s$
	3.6e−2	0.3	50	$0.5\,K_s$
	3.6e−2	0.3	60	$0.5\,K_s$
4	3.6e−2	0.2	35	K_s
	3.6e−2	0.2	35	$0.5\,K_s$
	3.6e−2	0.2	35	$0.25\,K_s$

5.2.4 Parameter Sensitivity Analysis of Slope Stability Under Rainfall

The test uses the Gardner model, as described by Eqs. (1.13)–(1.15). The slope soil layer is assumed to be homogeneous soil, the mathematical model is depicted in Fig. 5.1, its thickness is assumed to be 2 m, and the saturated and residual water contents are set to 0.46 and 0.1. The distribution characteristics of saturated permeability coefficient, desaturation coefficient α, slope angle, and rainfall q on porewater pressure are examined, and their influence on soil slope stability is discussed. The parameter settings of different conditions are shown in Table 5.5. For the slope stability analysis, the soil unit weight, effective cohesion, and effective internal friction angle are 19.9 kN/m³, 10 kPa, and 26°, respectively. The upper boundary is set as the flow boundary, the lower boundary is the groundwater level, its pressure head is set to 0, and the initial condition for removing the boundary points is $h(z, t = 0) = -z$ m.

5.2.4.1 Influence of Saturated Permeability Coefficient on Slope Stability

Figure 5.14 depicts the effect of three different saturated permeability coefficients on the pressure head distribution in Case 1. With the increase of the saturated permeability coefficient of the soil, the distribution of the pressure head moves faster, and

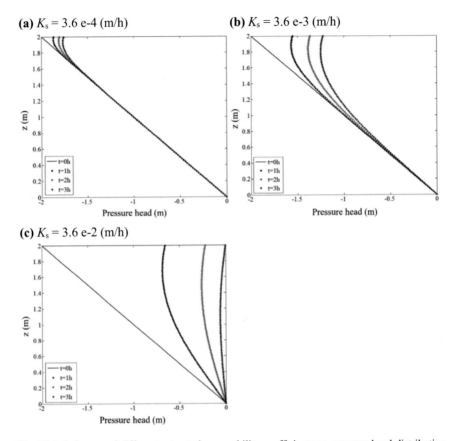

(a) $K_s = 3.6$ e-4 (m/h)   **(b)** $K_s = 3.6$ e-3 (m/h)

(c) $K_s = 3.6$ e-2 (m/h)

Fig. 5.14 Influence of different saturated permeability coefficients on pressure head distribution (Case 1)

with the increase of the rainfall time, the change of the pressure head is large. The numerical results illustrate that in the numerical simulation of rainfall infiltration for soil slope, the size of the saturated permeability coefficient can directly affect the speed of the pressure head distribution. Figure 5.15 indicates that the increase of the saturated permeability coefficient has a great influence on the stability of the shallow soil slope.

5.2.4.2 Influence of Desaturation Coefficient on Slope Stability

Figure 5.16 depicts the effect of different desaturation coefficients on the pressure head distribution in Case 2. As the desaturation coefficient increases, the pressure head distribution lags behind. At the same time, the curvature of the pressure head distribution becomes obvious with the increase of the desaturation coefficient.

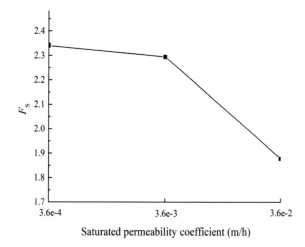

Fig. 5.15 Relationship between slope safety factor ($z = 1$ m) and saturated permeability coefficient

Figure 5.17 demonstrates that the slope safety factor decreases as the desaturation factor decreases.

5.2.4.3 Influence of Slope Angle on Slope Stability

Figure 5.18 represents the effect of different slope angles on the pressure head distribution in Case 3. As the slope angle increases, the pressure head distribution shifts gradually to the right. At the same time, the curvature of the pressure head distribution becomes obvious with the increase of the slope angle. Figure 5.19 shows that the slope angle has an inherent key effect on the slope safety coefficient, and the slope safety factor decreases continuously with the increase of the slope angle.

5.2.4.4 Influence of Rainfall on Slope Stability

Figure 5.20 depicts the effect of different rainfall on the pressure head distribution in Case 4. With the increase of rainfall, the migration speed of the pressure head distribution increases, which means that the soil matric suction dissipates faster. At the same time, the curvature of the pressure head distribution becomes obvious with the increase of rainfall. Figure 5.21 verifies that the slope safety factor decreases with increasing rainfall.

To sum up, in the analysis of rainfall slope stability, the slope angle has an inherent key role in the factor of safety, and the saturation permeability coefficient, desaturation coefficient, and rainfall all have a large influence on the distribution of pressure head. For the short-term rainfall period, when the desaturation coefficient is small and the saturated permeability coefficient is large, the larger the rainfall, the larger the rainwater infiltration depth, which will lead to the softening of the soil, the reduction

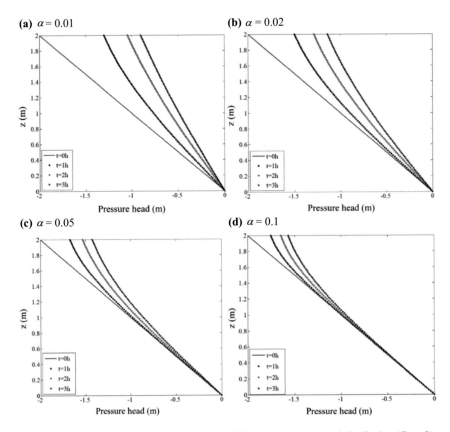

Fig. 5.16 Influence of different desaturation coefficients on pressure head distribution (Case 2)

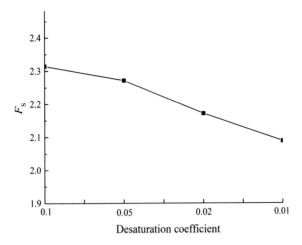

Fig. 5.17 Relationship between the slope safety factor ($z = 1$ m) and the desaturation factor

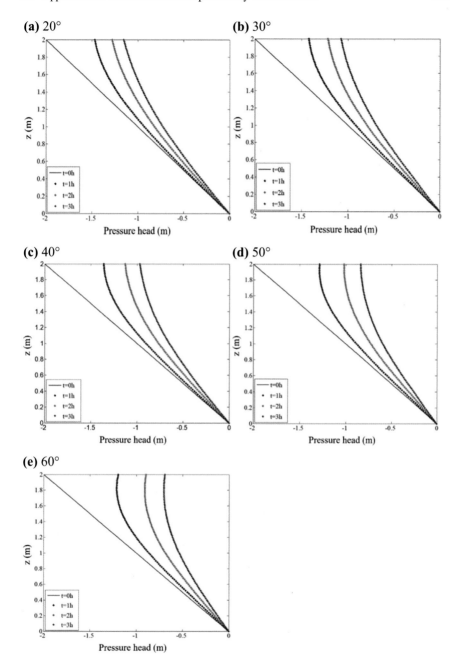

Fig. 5.18 Influence of different slope angles on pressure head distribution (Case 3)

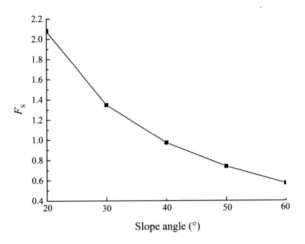

Fig. 5.19 Relationship between the slope safety factor ($z = 0$ m) and the slope angle

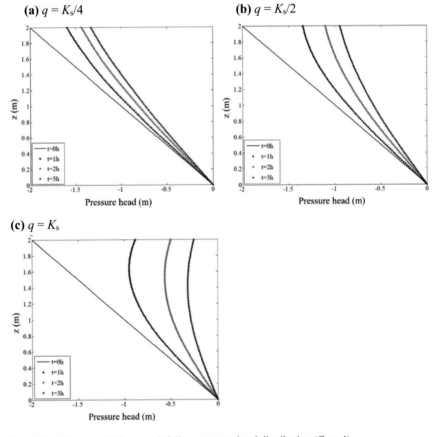

Fig. 5.20 Influence of different rainfall on pressure head distribution (Case 4)

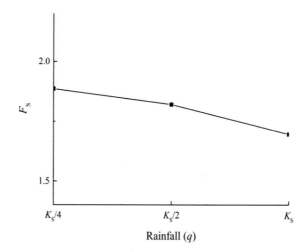

Fig. 5.21 Relationship between the slope safety factor ($z = 1$ m) and the rainfall

of the shear strength, and the shallow soil slope stability. As a result, factors such as desaturation coefficient, rainfall, permeability coefficient, and slope angle all affect the stability of unsaturated soil slopes.

5.3 Rainfall Landslide Case Study—Tung Chung Landslide

According to the improved method, a comparative study is carried out with the monitoring data of the pore-water pressure of a slope in Tung Chung, Hong Kong, to verify the application effect of the proposed method. The long-term rainfall monitoring in this area provides systematic data for investigating rainfall-induced landslides. Many researchers have conducted extensive and in-depth studies on landslides in the Lantau area, including field experiments, statistical analysis, and remote sensing interpretation (Zhang et al. 2016). The mathematical model of this application case is shown in Fig. 5.1.

The measured data of rainfall and pressure head at monitoring point SP3 are shown in Fig. 5.22. The change of pressure head is basically consistent with the change of rainfall, which better reflects the change of infiltration boundary. The monitoring data can be divided into three periods according to the time of the rainfall peak. Among them, the calibration period is from 0 to 192 h (0:00 on June 8, 2001 to 0:00 on June 16, 2001), and the Bayesian stochastic inversion of soil hydraulic parameters can be performed according to this stage (Yang et al. 2018). The model parameters and permeability coefficients of the two soil–water characteristic curves are shown in Table 5.6. After calibrating the soil hydraulic parameters by random back analysis, the error between the model calculated and measured values is very small (Yang et al. 2018). Numerical simulations are carried out for two rainfall periods in the verification period and compared with the monitoring data. The first period is 5 h (from 4:00 on June 27 to 9:00 on June 27), and the average rainfall is 1.4 cm/h. The second period is 5 h (from 1:00 on July 7 to 6 on July 7), and the average rainfall is 2.2 cm/h. The third period is 3 h (23:00 on July 15 to 2:00 on July 16), and the average rainfall is 2.4 cm/h.

The water pressure meter SP3 is buried at a depth of 200 cm, and the groundwater table depth is 250 cm. The thickness of the soil layer (L) is 250 cm, and β is assumed to be 35°. The initial conditions and boundary conditions are expressed as follows:

$$\left[K\frac{\partial h}{\partial z} + K\cos\beta \right]_{z=L} = q \tag{5.9}$$

$$h_{|z=0} = 0 \tag{5.10}$$

The space step is 2.5 cm, and the time step is 0.1 h. Figures 5.23 and 5.24 compare the results between the numerical solutions of the Richards' equation of the VGM and Gardner models and the measured values in the first period. It can be found that the Richards' equation described by the VGM model fits the measured values better. When $t = 5$ h, the relative error between the simulated and measured values is only 1.23%.

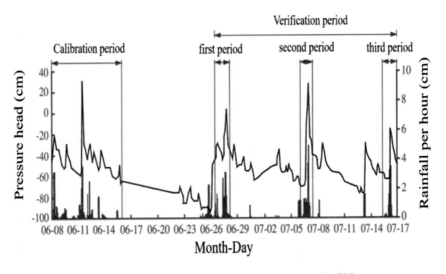

Fig. 5.22   Measured data of rainfall and pressure head at monitoring point SP3

Table 5.6   Soil–water characteristic curve and saturated permeability coefficient

Model	K_s (cm/h)	θ_s	θ_r	α	n
Gardner	3.33	0.38	0.10	3.09e−3	–
VGM	3.33	0.413	0.052	3.23e−3	1.58

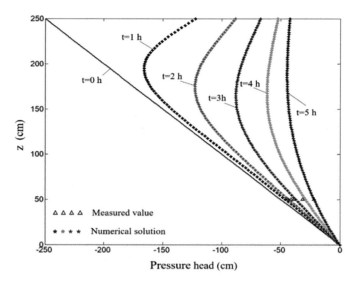

Fig. 5.23 Comparison between the numerical solution of Richards' equation for the VGM model and the measured value in the first period

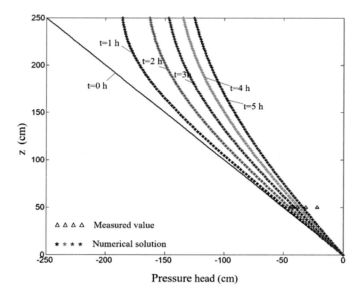

Fig. 5.24 Comparison between the numerical solution of Richards' equation for the Gardner model and the measured value in the first period

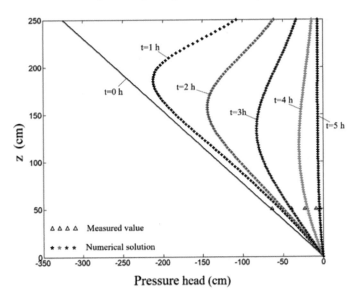

Fig. 5.25 Comparison between the numerical solution of Richards' equation for the VGM model and the measured value in the second period

Figures 5.25 and 5.26 represent the comparison results between the numerical solutions of the Richards' equation of the VGM and Gardner models and the measured values in the second period. Compared with the Gardner model, the fitting effect of the Richards' equation described by the VGM model and the measured values is still better. The relative error is 9.22% when $t = 5$ h.

Figures 5.27 and 5.28 compare the numerical solutions of the Richards' equation and measured values of the VGM and Gardner models in the third period. When $t = 3$ h, the relative error between the simulated value and the measured value computed by the two models is 3.54% and 3.67%, respectively. The numerical results of Richards' equation using the VGM model are closer to the measured values.

As shown in Table 5.7, in the three time periods, the overall root mean square error (RSE) and relative error (RE) of the VGM model and the measured data are both smaller than the values obtained by the Gardner model, which further indicates that the Richards' equation described by the VGM model has a better fitting effect with the measured data. Numerical results illustrate that the proposed method can well simulate the time-varying response of pressure head in the rainfall infiltration of unsaturated soil slopes and has a good application effect.

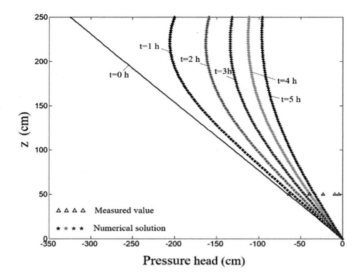

Fig. 5.26 Comparison between the numerical solution of Richards' equation for the Gardner model and the measured value in the second period

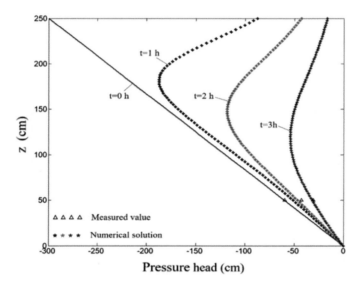

Fig. 5.27 Comparison between the numerical solution of Richards' equation for the VGM model and the measured value in the third period

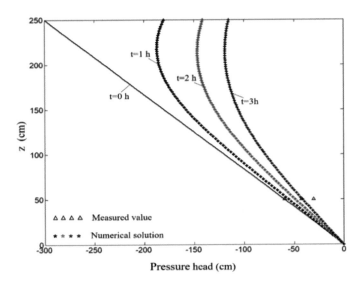

Fig. 5.28 Comparison between the numerical solution of Richards' equation for the Gardner model and the measured value in the third period

Table 5.7 Error between the simulated and the measured values during the validation period

Model	First period		Second period		Third period	
	RSE	RE (%)	RSE	RE (%)	RSE	RE (%)
Gardner	2.62e−1	18.87	3.44	229.8	1.79e−1	11.64
VGM	4.93e−2	4.33	6.25e−1	47.74	7.85e−2	5.42

5.4   Conclusions

This chapter studies the application of Chebyshev spectral method in slope stability analysis and the application of improved iterative methods P-CSIM and NTG-PI in shallow rainfall slope stability analysis. In order to verify the accuracy and practicability of the proposed improved method, a comparative study was carried out combining the monitoring data of slope in the Tung Chung area of Hong Kong and the Xiaoba landslide area. The conclusions are drawn as follows:

(1) The numerical results obtained by the CSM method are highly consistent with the transient analytical solution, which indicates that the proposed method is sufficiently accurate to handle transient infiltration problems associated with rainfall-induced landslides. The numerical solution obtained by the CSM method is introduced into the slope stability analysis to effectively evaluate the slope stability under rainfall conditions. The numerical solutions of the pressure head obtained by the improved methods P-CSIM and NTG-PI can effectively

analyze the shallow landslides caused by rainfall. Combined with the rainfall data, a mathematical model was established for the monitoring points, the evolution process of the pressure head of the slope was solved by an improved iterative method, and the stability analysis of the unsaturated soil slope was carried out. The results indicate that the numerical solution is consistent with the measured value, and the relative error is small, showing a good application prospect. The numerical solution to coupled water flow and deformation in two-layer unsaturated porous media is obtained using a finite element method.

(2) A conceptual model of two-layer unsaturated soils is established to analyze the rainfall infiltration process under different conditions. A simplified analysis of an infinite slope is used to compute the factor of safety as a function of the depth of wetting front, and special attention is paid to the hydrological response at the interface between two layers of unsaturated porous media.

References

Ali A, Huang JS, Lyamin AV, Sloan SW, Cassidy MJ (2014) Boundary effects of rainfall-induced landslides. Comput Geotech 61:341–354

Brooks RH, Corey AT (1964) Hydraulic properties of porous media and their relation to drainage design. Trans ASAE 7(1):26–28

Chen CY, Chen TC, Yu WH, Lin SC (2005) Analysis of time-varying rainfall infiltration induced landslide. Environ Geol 48:466–479

Cho SE, Lee SR (2001) Instability of unsaturated soil slopes due to infiltration. Comput Geotech 28:185–208

Conte E, Troncone A (2008) Soil layer response to pore pressure variations at the boundary. Geotechnique 58 (1):37–44

Collins BD, Znidarcic D (2004) Stability analyses of rainfall induced landslides. J Geotech Geoenviron 130(4):362–372

D'Aniello A, Cimorelli L, Cozzolino L (2019) The influence of soil stochastic heterogeneity and facility dimensions on stormwater infiltration facilities performance. Water Resour Manage 33: 2399–2415

Garcia E, Oka F, Kimoto S (2011) Numerical analysis of a one-dimensional infiltration problem in unsaturated soil by a seepage–deformation coupled method. Int J Numer Anal Meth Geomech 35(5):544–568

Iverson RM (2000) Landslide triggering by rain infiltration. Water Resour Res 36(7):1897–1910

Kim J, Jeong S, Park S, Sharma J (2004) Influence of rainfall-induced wetting on the stability of weathered soils slopes. Eng Geol 75:251–262

Liu CY, Ku CY, Xiao JE et al (2017) Numerical modeling of unsaturated layered soil for rainfall-induced shallow landslides. J Environ Eng Landsc Manag 25(4):329–341

Lu N, Likos WJ (2004) Unsaturated soil mechanics. Wiley

Ng CWW, Shi Q (1998) A numerical investigation of the stability of unsaturated soil slopes subjected to transient seepage. Comput Geotech 22(1):1–28

Masoudian MS, Afrapolic, MAH, Tasallotid A, Marshall AM (2019) A general framework for coupled hydro-mechanical modelling of rainfall induced in-stability in unsaturated slopes with multivariate random fields. Comput Geotech 115: 103162

Rahimi A, Rahardjo H, Leong EC (2010) Effect of hydraulic properties of soil on rainfall-induced slope failure. Eng Geol 114:135–143

Srivastava R, Yeh TCJ (1991) Analytical solutions for one-dimensional, transient infiltration toward the water table in homogeneous and layered soils. Water Resour Res 27(5): 753-762

Srivastava A, Kumari N, Maza M (2020) Hydrological response to agricultural land use heterogeneity using variable infiltration capacity model. Water Resour Manage 34:3779–3794

Wu LZ, Huang RQ, Xu Q, Zhang LM, Li HL (2015) Analysis of physical testing of rainfall-induced soil slope failures. Environ Earth Sci 73(12): 8519-8531

Wu LZ, Zhou Y, Sun P, Shi JS, Liu GG, Bai LY (2017) Laboratory character-ization of rainfall-induced loess slope failure. Catena 150:1–8

Wu LZ, Huang RQ, Li X (2020) Hydro-mechanical analysis of rainfall-induced landslides. Springer

Wu LZ, Cheng P, Zhou JT, Li SH (2022) Analytical solution of rainfall infiltration for vegetated slope in unsaturated soils considering hydro-mechanical effects. Catena. 206:105548

Xu QJ, Zhang LM (2010) The mechanism of a railway landslide caused by rainfall. Landslides 7(2):149–156

Yang HQ, Zhang LL, Li DQ (2018) Efficient method for probabilistic estimation of spatially varied hydraulic properties in a soil slope based on field responses: a Bayesian approach. Comput Geotech 102:262–272

Zha Y, Yang J, Shi L, Song X (2013) Simulating one-dimensional unsaturated flow in heterogeneous soils with water content-based Richards equation. Vadose Zone J 12(2):1–13

Zhan TLT, Jia GW, Chen YM, Fredlund DG, Li H (2013) An analytical solution for rainfall infiltration into an unsaturated infinite slope and its application to slope stability analysis. Int J Numer Anal Methods Geomach 37(12):1737–1760

Zhang LL, Li JH, Li X, Zhang J, Zhu H (2016) Rainfall-induced soil slope failure: stability analysis and probabilistic assessment. Shanghai Jiaotong University Press

Zhu SR, Wu LZ, Huang JS (2022) Application of an improved P(m)-SOR iteration method for flow in partially saturated soils. Comput Geosci 26(1):131–145

Zhuang JQ, Peng JB, Wang GH, Javed I, Wang Y, Li W (2018) Distribution and characteristics of landslide in Loess Plateau: a case study in Shaanxi province. Eng Geol 236:89–95

Zhang LL, Zhang LM, Tang WH (2005) Rainfall-induced slope failure consider-ing variability of soil properties. Geotechnique 55(2):183–188

Zhu SR, Wu LZ, Peng JB (2020) An improved Chebyshev semi-iterative method for simulating rainfall infiltration in unsaturated soils and its application to shallow landslides. J Hydrol 590:125157

Correction to: Rainfall Infiltration in Unsaturated Soil Slope Failure

Correction to:
L. Wu and J. Zhou, *Rainfall Infiltration in Unsaturated Soil*
Slope Failure, **SpringerBriefs in Applied Sciences**
and Technology, https://doi.org/10.1007/978-981-19-9737-2

Author provided new funder information "Natural Science Foundation Innovation and Development Foundation of Chongqing (No. CSTB2022NSCQ-LZX0044)" has been included in the acknowledgement.

The updated version of the book can be found at
https://doi.org/10.1007/978-981-19-9737-2

Printed in the United States
by Baker & Taylor Publisher Services